The Construction Industry
of Great Britain

Nothing endures but change
Heraclitus
c.540–480 BC

The Construction Industry of Great Britain

Second edition

EurIng, Roger C. Harvey, BSc(Eng), PhD, CEng, FICE, FIStructE, MRINA

The late lecturer in Civil Engineering, Queen Mary College, University of London and former Professor and Head of Civil Engineering, Sunderland Polytechnic

and

Allan Ashworth, MSc, ARICS

Former lecturer in Quantity Surveying, University of Salford and sometime Professor of Quantity Surveying, University of Technology, Malaysia

Laxton's

Laxton's
An imprint of Butterworth-Heinemann
Linacre House, Jordan Hill, Oxford OX2 8DP

A division of Reed Educational and Professional Publishing Ltd

ᴿ A member of the Reed Elsevier plc group

OXFORD BOSTON JOHANNESBURG
MELBOURNE NEW DELHI SINGAPORE

First published 1993
Second edition 1997

British Library Cataloguing in Publication Data
A catalogue record for this book is available from the British Library

ISBN 0 7506 3656 4

Composition by Hope Services (Abingdon) Ltd
Printed and bound in Great Britain

Contents

Preface to the second edition

When we prepared the first edition of *The Construction Industry of Great Britain* in 1992–93 the industry was in the middle of one of its worst recessions. The caption to Figure 1.1 continues to provide a poignant reminder of the good and bad fortunes of the industry. This also seeks to emphasize the fragile nature of the economic recovery in the country in general but particularly in this industry. Various reports continue to show, for example, that house prices are in the doldrums (Halifax Building Society) and that housing turnover has stagnated (Inland Revenue). Industry pundits accept that all is not well but optimistically point to better times ahead as the millennium is approached.

The response to the first edition from colleagues and students, on a wide range of different courses, has been sufficient to encourage us to write a second edition of this book. This has been done to update the data on which the book is based, to include issues raised through recent reports such as *Constructing the Team* (Latham) and to add new material that is now of importance. In addition, we have made general and textual revisions that were necessary with changes that have occurred in the industry.

The common issues of the industry largely remain. There is no end in sight to the boom and bust mentality of the industry. Whilst government remains a major client its involvement in the construction industry is declining, due in part to the privatization of nationalized industries. Recently, there has been encouragement to carry out many public works projects through the use of the Private Finance Initiative (PFI) and the use of National Lottery funds. The separation of design from construction is still a popular method used to obtain construction projects and, whilst there are those protagonists who see the traditional procurement of buildings and engineering works as outdated, there may be some suggestion that through the introduction of project management and single point responsibility it is not the designer who may disappear from the scene but the major British constructor. Like the British motor car industry the latter are now fewer in number and in the hands of more foreign ownership. An eminent professor suggested recently that the artificial division between building and civil engineering will have to disappear as the industry becomes more involved in Europe, the USA and the rest of the world.

Amidst all of this the construction industry of Great Britain continues to provide useful and interesting employment, innovation in design and construction and buildings and structures around the world which are a tribute to all of those involved. It is at times like this that the dynamic and changeable nature of the construction industry of Great Britain is evident.

We continue to be grateful to those firms, organizations and individuals that have allowed us to use their information and data. Due acknowledgement is provided throughout the text to the different sources that have been used in the preparation of the second edition of *The Construction Industry of Great Britain*.

Sadly Roger Harvey died just before the book was published and was thus not able to enjoy the fruits of his work.

Allan Ashworth
Roger C. Harvey
1997

Preface to the first edition

The construction industry of Great Britain is unique. It has literally built Britain, and its monuments are around for all to see. It is one of few industries able to produce goods that increase in value over time, unlike the majority of other products that begin to depreciate immediately from the time of purchase. The construction industry, in addition to providing investment, offers a wide range of employment to all stratas of society. However, it is an industry which is highly susceptible to the booms and slumps in the economy and to the stop–go policies of government. A less exaggerated increase or decrease in workloads would be preferable to encourage longer term and continuous investment to take place. This might end the boom-and-bust philosophy that has dogged the industry for the better part of this century. The industry undertakes much of its activities, not in its own workshops, but on the clients' premises. It has traditionally separated the design from the production of its buildings and structures and, unlike other countries has encouraged a fragmentation of professional activities. There are indications that the education of its different professionals is moving towards a more common base and this should, at least, provide a better integration of the different professional roles. In comparison with some countries, the industry has a reputation for not performing well in respect of the quality of the product, the time required for completion and the overall costs involved. The reasons for this include shortcomings in management and supervision, fragmentation of the design and production activities, a workforce which is sometimes poorly skilled and motivated, and a dearth of research and development. Amidst all of this the construction industry of Great Britain has produced many notable buildings and engineering structures both at home and overseas. Some are amongst the finest in the world. This book examines the industry and attempts to provide a picture of its activities.

The book is intended to be of use to students studying on a range of different courses associated with the construction industry and for other students who may need to gain some knowledge of the industry's activities. It is expected that practitioners will find it an interesting study of the industry in which they work.

We are grateful to those firms and organizations which have provided or allowed us to use their information, and due acknowledgement is provided.

Allan Ashworth
Roger C Harvey
1993

Background

Great Britain comprises England, Scotland and Wales. It is about the same size as New Zealand or Uganda, half the size of France and six times the size of the Netherlands, Switzerland or Denmark. It is just under 1000 km from the south coast to the extreme north of Scotland and just under 500 km across its widest part. No place in Britain is more than 120 km from tidal water. The prime meridian of 0 degrees passes through the old observatory at Greenwich (London).

The population of Great Britain in 1996 was 56.55 million. This compares with 9 million at the end of the seventeenth century and almost 38 million at the turn of this century. Projections suggest that by the year 2000 the population will have risen to about 57 million. The population is the nineteenth largest in the world and is comparable with that of France (57 million), Italy (57 million), Turkey (59 million), Egypt (58 million) and Thailand (58 million). In Britain, 19% of the population is under 15, 65% between the ages of 15 and 64 and 16% is over 65 years of age. The population grew steadily up to 1971, but has grown more slowly since then. The population decline in the mid-1970s occurred for the first time since records began, other than during wartime. However, the age structure has changed with a lower proportion of children aged under 15 and a higher proportion of adults aged 65 and over. By the early part of the next century 20% of the population will be over 65. Whilst more males than females are born, their greater infant mortality and shorter life span of 72 years against 78 years, result in the latter outnumbering males by approximately 5%.

The population density is in the order of 236 inhabitants per square kilometre (65 in Scotland and 365 in England). This is about the same as Germany and India but well above the European Community average of 170 inhabitants in 1988. The population density of Japan and Belgium is about one-third higher, whilst that of France is less than half and the USA only one-tenth. Since the nineteenth century there has been a trend to move away from the congested urban centres into the suburbs. There has also been a geographical redistribution of the population from Scotland and northern England to East Anglia and the south.

The way of life of the people in Britain has been changing rapidly in the second half of the twentieth century. The underlying causes include lower birth rates, longevity of life, widening educational opportunities, technical progress and higher standards of living.

Acknowledgements

Statistics are reproduced by permission of the following: Housing and construction statistics, by the Department of the Environment and as Crown Copyright by the Controller of Her Majesty's Stationery Office; Annual Abstract of Statistics, Regional Trends, by the Central Statistical Office; Quality in Traditional housing, How Flexible is Construction; Faster building for Industry, Strategy for Construction R&D, Construction for Industrial Recovery, Blackspot Construction, Annual Review of Government Funded R&D, as Crown copyright by the Controller of Her Majesty's Stationery Office; The Top Fifty Contractors (Annual Survey), Building without Conflict, Sizing up Trouble, by Building Magazine, Britain 1990: An Official Handbook, by the Central Office of Information and as Crown Copyright by the Controller of Her Majesty's Stationery Office; patent information by the British Library; information research funding bodies by Thomas Telford Ltd; research information on the Building Research Establishment.

Quotations are reproduced by permission of the Building Research Establishment.

Figures are reproduced by permission of the following: Figure 1.1, Shepherd Homes; Figure 2.1, Stanhope Properties; Figure 3.4, Wimpey PLC; Figures 3.5, 9.1, 12.1, Bovis PLC; Figure 5.1, the Controller of Her Majesty's Stationery Office; Figure 5.2, the Kajima Corporation and the Penta-Ocean Construction Company Ltd; Figure 7.1, Costain PLC; Figure 11.4, the Health and Safety Executive and as Crown Copyright by the Controller of Her Majesty's Stationery Office; Figure 16.3, NCL Stewart Scott Ltd, Consulting Engineers.

Many persons have discussed, and commented on, ideas and have given information. We are indebted to them all.

A.A.
R.C.H.

1

Economic significance of the construction industry

If building is to be looked upon as a tap which can be turned on and off for economic reasons, then efficiency cannot be expected.

Ministry of Public Building and Works, 1950

Figure 1.1 *When new housing begins to be constructed the economy is moving out of recession*

The construction industry

The construction industry has literally built Great Britain. Its activities are concerned with the planning, regulation, design, manufacture, construction and maintenance of buildings and other structures. The industry embraces the sectors of building and civil engineering and also includes the process-plant industry, although the demarcation between these different areas is blurred. Construction work includes a wide variety of different activities in respect of the size and type of projects which are undertaken and the professional and trade skills that are required. Projects can vary from work worth a few hundred pounds undertaken by jobbing builders, to major schemes costing several million pounds and projects such as the Channel Tunnel which is an international joint venture, estimated to cost over £10bn. Whilst the principles of execution are similar, the scale, complexity and intricacy vary enormously. The total value of capital stock of works which have been constructed in Britain are estimated to be worth £2500bn. The construction industry of Great Britain is also responsible for a significant amount of work undertaken overseas on behalf of British consultants and contractors. The latter typically represents between 10 and 15% of the annual turnover of the major contractors.

The construction industry has characteristics which separate it from all other industries. These are:

1   The physical nature of the product.
2   The product is normally manufactured on the client's premises, i.e. the construction site.
3   Many of its projects are one-off designs and lack any prototype model being available.
4   The arrangement of the industry, where design has normally been separate from construction.
5   The organization of the construction process.
6   The methods used for price determination.

The final product is often large and expensive. It often represents a client's largest single capital outlay. The project is sometimes required over a large geographical area, such as road schemes. Buildings and other structures are, for their most part, bespoke designed and manufactured as products to suit the individual needs of each customer, although there is provision for repetitive and speculative work, particularly in the case of house building. The nature of the work also means that an individual project can often represent a large proportion of the turnover of a single contractor in any year, causing little continuity in the production functions.

The output of the economy

Production is expressed in terms of value added, or the added value of output of net inputs, so as to avoid double counting when the sales of the different units are added together. It can be broadly subdivided as shown in Table 1.1.

Energy was the most rapidly expanding sector in the early 1980s, because of the rising North Sea oil production, but this has now reversed as oil production has passed its peak. The manufacturing

Table 1.1   Employment sectors

Sector	%
Primary [agriculture/energy]	9
Secondary [manufacturing/construction]	30
Services	61
Total	100

sector has continued to decline over many years, particularly in the so-called heavy industries. There is of course fluctuation in this sector and whilst much of engineering is in terminal decline, electricals and chemicals are continuing to expand. The output of the construction industry ebbs and flows as a key indicator of the economy. However, it does rely to a large extent on the good fortunes of other industries or sectors for much of its work. The service sector which includes finance, communications, distribution, hotels and catering are growth areas.

Table 1.2   Typical percentage shares of GDP (1988) and projected employment change 1990–2000 [1, 2]

Sector	GDP %	Projected employment change 1990–2000
Agriculture/energy	7	–20
Manufacturing	24	–25
Construction	6	+3
Distribution, hotels/catering	14	+2
Financial services	20	+12
Others	29	+16
Total	100	+3

The value of goods and services produced for final use in consumption, capital expenditure and exports is called the gross domestic product [GDP]. Construction accounts for about 6% of GDP as shown in Table 1.2. The gross national product [GNP] adds net property income from abroad to the GDP. Britain has a total GDP of $1 trillion, which is 7% of the G7 (the seven countries in the world economic league) total. It has the world's sixth largest GDP after, USA, Japan,

Germany, France and Italy. However, it ranks lower in terms of the rate of growth of GDP in real terms and GNP per head of population. It had the world's seventeenth largest GNP per head in 1988. The highest annual average growth rate in GDP achieved over a decade was in the 1960s, at about 2.9% in real terms. It was negative at the start of the 1980s, averaged 3.3% between 1982 and 88, but due to the impact of high interest rates declined in the early 1990s. The recession since the summer of 1990 is the longest since the 1930s.

The workforce in employment, which included those on government training schemes, in 1990 was about 26.2m or 47% of the total population. The distribution of these is shown in Table 1.3. During the 1980s the workforce increased by over 1.8m people but employees in employment remained about the same at about 21m. Self-employment, a culture which developed during this period, accounted for about 1.3m extra jobs and these also increased in proportion from 7.5% to 11.6% of the workforce. Self-employment in the construction industry at least increased in line with this trend. In addition, almost half a million people were on government training schemes. Unemployment peaked in 1986 at 3m but this temporarily declined up to the end of the decade but was to rise again to over 10% of the workforce by the early 1990s. The projected employment change by the different industry groups is shown in Table 1.2.

Employment rose substantially in the second half of the 1980s, with 2.9m extra jobs between March 1983 and March 1989. This represented a larger rise than any other country in Europe. However, due to the recession of the early 1990s about 1000 jobs per day were lost. Many of these were in construction, as the industry faced one of its worst recessions since the 1930s.

Male employees account for 56% of the workforce in employment and females 44%, nearly half of the latter work part-time. The construction industry employs relatively few women even amongst the many professions. Unemployed claimants stood at 1.6m in March 1990, although this increased to over 2.5m by one year later. In July 1990 the adult unemployment rate, seasonally adjusted, was 5.7%. Table 1.4 provides a comparison of employees in the construction industry against the other major market sectors. After 1990 these numbers began to decline as unemployment began to rise. The primary sector is in long-term decline as are many of the manufacturing industries. Growth has occurred in the service sector, but by the early 1990s this sector had also begun to show signs of stagnation as banking, insurance and other financial employers began to reduce the numbers in their workforces.

Unemployment rose again during the early 1990s, peaking in 1993

Table 1.3 Distribution of the workforce [3]

	1980	1985	1986	1987	1988	1989	1990	1991	1992	1993	1994	1995	1980	1990	1995
	thousand												*percentages*		
Employees	22 432	20 910	20 876	21 081	21 748	22 143	22 334	21 707	21 359	21 039	21 081	21 321	86.5	80.4	78.1
Self-employed	1 950	2 550	2 567	2 801	2 926	3 182	3 222	3 306	3 136	3 098	3 206	3 264	7.5	11.6	12.0
HM Forces	323	326	322	319	316	308	303	297	290	271	250	230	1.3	1.1	0.8
Govt training	–	168	218	303	335	452	412	333	307	295	286	247	–	1.4	0.9
Total	24 705	23 954	23 983	24 504	25 325	26 085	26 271	25 643	25 092	24 703	24 823	25 062	95.3	94.5	91.8
Unemployed	1 214	2 907	2 998	2 716	2 185	1 685	1 522	2 203	2 630	2 817	2 547	2 226	4.7	5.5	8.2
Workforce	25 919	26 861	26 981	27 220	27 510	27 770	27 793	27 846	27 722	27 520	27 370	27 288	100.0	100.0	100.0

Table 1.4 Employees in employment [3]

Industry	1985	1986	1987	1988	1989	1990	1991	1992	1993	1994	1995
					thousands						
Primary sector	903	846	801	770	736	722	697	631	552	483	472
Manufacturing	5 244	5 112	5 070	5 097	5 090	5 054	4 136	3 948	3 825	3 838	3 875
Construction	994	964	963	1 021	1 056	1 061	1 025	925	840	848	814
Services	13 769	13 954	14 247	14 860	15 261	15 497	15 849	15 855	15 822	15 912	16 160
Total	20 910	20 876	21 081	21 748	22 143	22 334	21 707	21 359	21 039	21 081	21 321

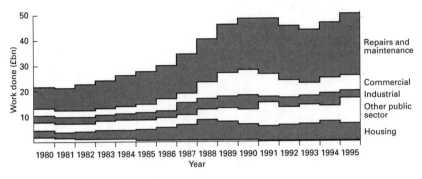

Figure 1.2   *Value of output [4]*

at 10.2%. This was the highest recorded figure since 1986, when unemployment exceed 11%. These figures would have been rather worse had it not been for early retirement schemes (retired people are not included in the statistics), staying on rates in schools and in further education and the huge expansion of higher education.

Importance of the construction industry

The importance of the construction industry in the economy of Britain (and indeed any country) is due to the following factors. It experienced rapid growth in the late 1980s, with a total value of output in excess of £55bn by 1990. This output included work done by contractors, including estimates of unrecorded output by small firms, self-employed workers and direct labour departments. This was in accordance with the 1980 Standard Industrial Classification.

Scale of the construction industry

The construction industry accounts for about 6% of GDP. It provides over half of the fixed capital investment (e.g. ships, vehicles, aircraft, plant, etc.) in Britain. It experienced rapid growth in the late 1980s with the total value of output reaching almost £45bn by 1990. However, the recession throughout the country at the start of the 1990s had severe repercussions on the construction industry with output falling in real terms. The impact was more serious than the figures suggest due, in part, to the volume of work which was already in progress, and to a number of major projects which were under construction such as the Channel Tunnel project and the stadia required for the World Student games in Sheffield.

Table 1.5 Value of output [4]

	Value of output £1m at current prices							
	1980	1985	1900	1991	1992	1993	1994	1995
New housing								
Public	1 711	843	934	793	1 243	1 415	1 671	1 656
Private	2 585	3 797	5 746	5 003	4 841	5 213	5 746	5 470
Other new work								
Infrastructure	incl	2 254	4 965	6 062	5 716	5 544	5 149	5 647
Public	3 524	2 611	4 414	4 142	4 181	4 045	4 384	4 650
Industrial	2 806	2 159	3 394	2 622	2 234	2 208	2 489	2 996
Commercial	2 430	3 642	11 310	9 103	6 600	5 131	5 648	6 208
Repair/maintenance	8 997	14 358	24 544	23 389	22 658	22 767	24 353	25 900
Totals	22 053	29 664	55 307	51 114	47 473	46 323	49 440	52 527
Percentage new	59	52	56	54	52	51	51	51

In the first edition of *The Construction Industry of Great Britain* (1993) it was suggested that new work typically represented about 60% of the total construction output. Based upon a revised classification of data as indicated above, the value of new work is nearer 50% (Table 1.5). The figures exclude the DIY market, which is estimated to be worth £20bn and has seen a rapid expansion in recent years. This continues to be rated as the most important hobby of men.

Table 1.5 shows clearly the rise in workload in the late 1980s and the decline from the peak performance of 1990 through to the present day, remembering that these are current prices. The total output compares poorly with that of Britain's largest company, BP, whose turnover in 1995 was worth in excess of £30bn. Nevertheless the industry is substantial, offering direct employment to almost two million people (see Table 1.7) and to many others in supporting occupations. It continues to be the fourth largest construction industry in Europe representing about 9% of the total output (see Table 1.9). However, it is dwarfed by the German construction industry that has a 32% share of Europe. This emphasizes the massive redevelopments currently being carried out in the former East Germany. It is also superseded by France (14%) and Italy (12%) but larger than that of Spain (7%).

The largest variation in output occurs in the private commercial sector. This peaked in 1990, during which time the massive developments were under construction at Canary Wharf in Docklands. In 1990, it was worth £11.3bn representing 20% of the total output. By 1994 this had fallen to £5.1bn (11%), although there are indications that this sector is now beginning to expand (£6.2bn in 1995). Private industrial buildings, by comparison typically represent about 5% of output. The peak in private house building occurred earlier in 1988, signalling a boom in the economy (Figure 1.1). In 1988, it was worth £7.7bn (17%) of the output. By 1992, the height of the recession in the construction industry this had fallen to £4.8bn (10%). Public sector housing has slowly been growing in the 1990s to in excess of £1.5bn (3%). In 1991, it represented half of this amount.

In 1980 the public sector share of the total output of new work was £5.7bn (39%). This increased to £10.3bn by 1990, although it fell as a proportion (33%). By 1995 it had increased to £11.95bn (45%). A large proportion of this was in infrastructure works representing about 50% of all new public sector work, 20% of the total output of new work and over 10% of the total output including repairs and maintenance.

The public sector clients can be grouped under the classifications of central government, two-tier local government, i.e. county and borough authorities, and public corporations, although many listed under this heading have now become private sector corporations. Construction work in the public sector is mainly undertaken for social or political reasons. The private sector clients include owner-occupiers, investment groups, speculative developers and clients such as churches and leisure organizations. The general motivation for their developments is investment, although some do undertake work of a charitable nature.

Regional variation

There is an increased economic dominance in activity in the south-east of England compared with the other regions. This accounted for over 40% of the construction industry's output in the late 1980s compared with over 30% in the early 1970s (Table 1.6, Figure 1.3). This represents a distorted picture when compared with the population distribution. The rise is partially accounted for by the disproportionate increase in the demand for office space over that period. Ironically, land on the M4 corridor that was for sale at the height of the construction boom of the late 1980s was selling for a fifth of its price by the end of 1991. The four most northerly regions, known commonly as the north of England and Scotland now account for less than 30% of the nation's construction work. This is in decline compared with over 35% of construction in the early 1970s. The regions at the extremities from London; Scotland, Wales and the south-west also indicate a decline in the share of construction activity down by a fifth from 25% to 20% of the national share.

Table 1.6   Regional variation [5]

	1971	1975	1984	1988	1990	1994
North	242	443	1 145	1 717	2 209	2 139
Yorks/Humberside	274	484	1 972	3 132	4 093	3 703
East Midlands	273	409	1 538	2 731	3 283	3 301
East Anglia	155	216	1 030	1 985	2 098	1 977
South East	1 438	1 913	9 330	16 526	20 225	15598
South West	278	408	2 076	3 792	4 386	4 055
West Midlands	333	491	2 077	3 332	4 207	4 039
North West	448	634	2 325	3 526	4 687	4 576
Wales	220	443	992	1 561	2 097	2 172
Scotland	409	839	2 326	3 234	4 250	4 310

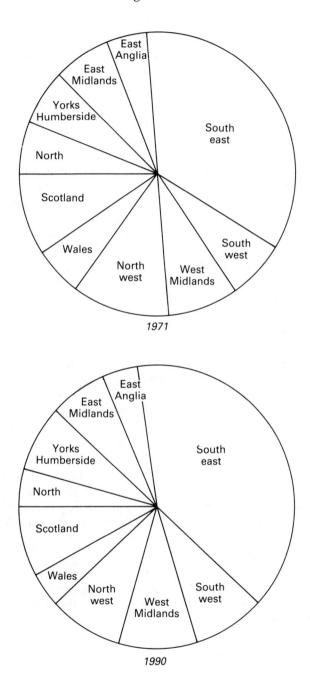

Figure 1.3   *Regional output of work undertaken by contractors [5]*

Employment

Table 1.7 are estimates from returns to the Department of Employment from contractors and public authorities. The estimates of employees who are not on the register are based upon the Central Statistical Office's Labour Force Survey. These figures exclude those employed in consulting firms and private practice. Those employed in that sector may boost the figures by as many as 250 000 up to in excess of 1.6m people. There are also a large number of others who are employed indirectly with materials and component manufacturers, plant and vehicle builders. In addition there is a whole range of secondary employment that relies upon a prosperous construction industry. These directly include lawyers and accountants, furniture manufacturers and removers, retail outlets, etc. and indirectly virtually every trade and profession.

The industry currently accounts directly for about 4.5% of the employed labour force. This figure can widely fluctuate, as Table 1.7 indicates and may be as much as 25% higher in times of boom in the industry. Within these figures there are about 700 000 operatives of their different kinds, although due to the changing nature of employment this figure declined steadily throughout the 1980s. During the same period of time the numbers of self-employed almost doubled and then declined at the start of the recession in the early 1990s. There are a further 300 000 employed in administrative, professional, technical and clerical occupations. The numbers of people employed in the repair and maintenance sector alone, is considerably greater than those in agriculture, coal mining, shipbuilding and many other traditional but declining industries.

The figures indicate that the increase from 1.6m individuals in 1980 to the peak in 1990 of 1.8m. By 1994 employment had declined by almost half a million. The numbers of employees in the public sector, both for operatives and APTC staff fell by over 50% between 1980 and 1994.

Investment

The construction industry produces goods which are predominantly of an investment nature. Its products are not wanted for their own sake but for the goods or services which they can help to create.

Government

Government is a large client of the construction industry (Figure 1.4, Table 1.8), even though as a whole the sector is shrinking due to the

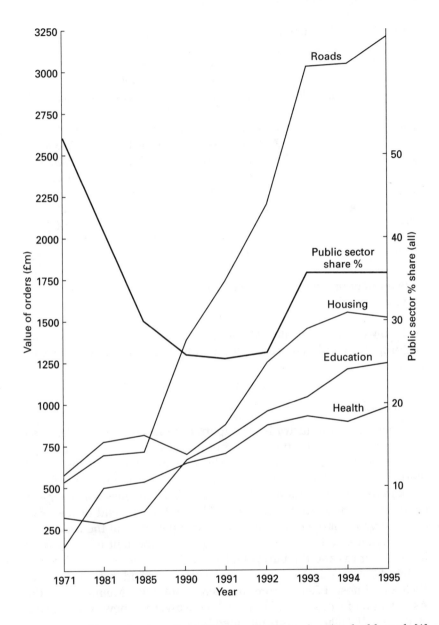

Figure 1.4 *New orders from the public sector; housing, education, health, roads [4] and public percentage share*

Table 1.7   Employment in the construction industry [4]

	Operatives	APTC*	Not Registered	Self-employed	Totals
	Thousands	Thousands	Thousands	Thousands	Thousands
1980	979	345	0	310	1634
1985	725	297	64	470	1556
1990	710	332	76	715	1833
1991	636	314	91	657	1698
1992	549	277	98	597	1521
1993	492	254	94	571	1411
1994	451	240	89	604	1384

*Administrative, professional and technical staff

large number of departments being privatized, and to a lower level of activity generally in this sector. In the early 1970s public building and works accounted for over 50% of the construction industry's work-load. However, by the late 1980s, this had fallen to less than 25%, and this proportion only recovered slightly due to recession in the private sector.

International

The construction industry is an important industry worldwide. Even in the poorer countries the net output of construction as a percentage of GNP is between 3% and 6%. The industry works in the world market (see Chapter 16) with many British firms being responsible for work in developing countries. In 1988 British companies won overseas contracts which were worth over £2.3bn and, at the end of the year, the total value of overseas work either completed or in the process of construction stood at £5.9bn. The industry's important markets were in the Americas and the European Union. British professionals such as architects, consulting engineers and surveyors had net earnings in 1988 of almost £700m from overseas contracts. Members of the Association of Consulting Engineers were awarded new work valued at almost £10 000m. Projects included [6]:

• Greater Cairo Wastewater Project £1000m
• Baghdad Metro £4000m
• Man-made river project in Libya £2300m
• Pipeline project in Nigeria £500m

Table 1.8 New orders from the public sector [4]

	New orders public sector [£m]											
	1971	1975	1980	1985	1986	1987	1988	1989	1990	1993	1994	1995
Housing	520	1 404	758	734	772	903	882	872	683	1 405	1 662	1 650
Education	306	335	330	342	364	369	501	611	673	1 080	1 187	1 246
Health	184	205	320	491	430	492	712	824	663	869	817	958
Roads	257	386	562	802	863	894	1 029	1 181	1 351	3 063	3 067	3 398
Other	853	1 368	2 037	2 243	2 485	2 758	2 874	3 588	2 460	1 917	2 163	2 238
Totals												
Public	2 120	3 698	4 007	4 612	4 914	5 416	5 998	7 076	5 830	8 334	8 896	9 490
All	4 071	6 279	10 115	15 343	17 108	22 119	26 299	27 142	22 492	23 249	24 786	26 346
Ratio (%) (Public: all)	52	59	40	30	29	24	23	26	26	36	36	36

- Housing in California £200m
- Road tunnel in Hong Kong £150m
- Highway in Turkey £140m

The export interests of the various sectors of the British construction industry are promoted by a number of different groups such as the Export Group for the Construction Industries, the British Consultants Bureau (which include architects, engineers and surveyors) and those who represent the materials suppliers, product manufacturers, construction plant and equipment firms.

Europe

The principal characteristics of Western Europe may be summarized as follows. The countries have a relatively small land area with great physical diversity. There are many nation states (Table 1.9) that are separated by historical precedent and language barriers. The countries are generally wealthy with relatively dense populations. They are industrialized and have an urbanized way of life. They are well watered environments with adequate fertile lands and a temperate cli-

Table 1.9   Construction output (£bn) country by country at constant 1994 prices and percentage changes

Country	1994	% change 1990–94	1996	% change 1994–96	2000	% change 1996–2000
Germany	181.6	5.7	185.2	1.0	202.8	2.3
France	81.8	−2.3	84.8	1.8	90.1	1.5
Italy	70.9	−2.2	73.0	1.4	79.4	2.1
UK	53.3	−2.4	53.4	0.1	55.5	1.0
Spain	38.5	−2.0	40.7	2.8	44.1	2.1
Netherlands	24.3	0.4	25.4	2.2	26.6	1.2
Switzerland	26.5	−2.0	25.6	−1.6	27.1	1.4
Belgium	18.5	1.4	19.3	2.2	21.0	2.1
Austria	24.4	3.6	25.5	2.2	26.8	1.3
Sweden	15.8	−3.4	17.0	4.1	18.8	2.5
Denmark	10.8	−3.1	12.6	7.6	14.1	3.0
Finland	7.6	−12.3	8.4	5.6	11.0	7.0
Norway	8.3	−1.1	9.2	5.5	9.0	−0.5
Portugal	7.4	2.0	8.4	6.7	10.4	5.6
Total	569.7	0.2	588.5	1.6	636.7	2.0

mate which supports a commercial system of agriculture. There are four physiographic regions and these underline the diversity. On the north-western rim lie the Atlantic uplands and Scandinavian mountains. The single most significant physical region is the north-west European plain, an extensive lowland stretching from the Bay of Biscay across France into the Rhinelands. A complex central region of mountains, valleys and plateaux lies from the Massif Central across the Rhine uplands. This is succeeded to the south by the Alpine system, with peaks exceeding 400 m, which stretches across the Mediterranean coastlands from Spain, Italy and Greece.

The population of Europe is approaching 300m which includes 57m living in Great Britain (20%). The population density is greatest in the Netherlands (347 people per square kilometre and least in France (50 people per square kilometre) compared with 236 in Great Britain. There are high densities of population recorded in the UK, Germany and Italy, but there are considerable regional variations. Whilst work in agriculture has declined to less than 5% in most of the European countries, it is much larger in the southern part of Europe. The percentages of the population described as urbanized vary from as many as 96% in Belgium to 31% in Portugal. The comparable figure for the UK is 92%.

The output of the construction industry in 1994 was worth almost £570bn (Tables 1.9 and 1.10). This is expected to increase to almost £640bn by the end of the century. Refurbishment currently represents 31% of the total output. This is a considerably lower percentage than that for the UK alone (Table 1.5). Civil engineering in Europe is worth £123bn or 22% of the total output. This is comparative to the proportions in the UK (Table 1.5).

Table 1.10   European construction output by sector and forecast percentage change until year 2000 (Source: Euroconstruct)

| Activity | 1994 (£bn) | % change at constant 1994 prices | | | |
		1994	1995	1996	1997–2000
New housing	149.9	8.8	2.3	−2.2	1.7
Private non-housing	86.6	−2.4	1.9	4.0	2.5
Public non-housing	31.5	0.1	−0.2	0.7	0.9
Total building	268.0	3.9	1.9	0.1	1.9
Civil engineering	122.9	0.6	1.7	2.4	2.3
Refurbishment	178.7	2.3	2.1	2.5	1.9
Total	569.6	2.7	1.9	1.4	2.0

By far the largest market is that of Germany, worth about one-third of the total output but which grew by a modest output of 1% between 1994 and 1996 and expected to grow 2.3% between 1996 and 2000. The big four (Germany, France, Italy and the UK) have markets worth almost £390bn or almost 70% of the total output.

Europe's construction markets face five years of steady, if modest growth until the millennium, according to the continent's leading forecaster, Euroconstruct. The forecaster attributes the sluggishness of the recovery, the average growth rate will be only 2% between 1994 and 2000, to the poor prospects for housing. This sector which accounts for more than a quarter (26%) of all European construction, faces a 2% decline in 1996 to be followed by a small increase of 1.7% over the following four years. A stronger growth is expected in the industrial and commercial sectors of 4% and 2.5% respectively.

Between 1994 and 1996, the UK market grew only 0.1%, the second lowest of the 14 countries. During the following four years the UK is also expected to under-perform all other European markets, except Norway, with just a 1% growth. The fortunes of Germany appear to rest on the revival of its housing sector, currently the subject of heavy government stimulation.

In Eastern Europe, the largest market is Poland, which is expected to grow by 5.5% a year in the second half of the 1990s. Euroconstruct forecasts that output in the year 2000 will be more than £14bn, 47% higher than in 1991. Since the late 1980s, construction activity has been restricted to housing, industrial and a few prestige commercial projects, mainly funded by foreign investors.

The economy

Throughout the 1980s the British economy grew at a rate of 3% per annum, the longest growth pattern since the end of the Second World War. Investment, employment, exports and productivity all increased. However, inflation began to rise towards the end of the decade and whilst unemployment had fallen by almost a half throughout the decade, by 1991 this had increased to 2.2m, with forecasts above 3m, before it would decline again. Youth unemployment, which had been a particular problem in the early 1980s, was forecast to disappear due to the demographic decline and increasing employment opportunities. This was not to be. The early part of the 1990s resulted in a recession throughout the economy and particularly in the construction industry. Whilst measures were taken by government to reduce interest rates and inflation, unemployment remained at a high level (Table 1.3). Confidence was beginning to return slowly due partially to economic stability, improved world markets and levels of demand that had been absent in previous years.

In managing an economy a government sets out to achieve:

- The ability to pay its way abroad by balancing the payments.
- An acceptable level of employment of resources, particularly people.
- An increase in the amount of goods and services produced and consumed, leading to a rise in the standard of living.
- The control of inflation.

The effects of changes in output, employment, incomes or demand in the construction industry have repercussions in other sectors of the economy through a knock-on effect. Thus a decline in construction will have an adverse effect upon other activities and industries in a market economy. Government economic policy can have the effect of stimulating or depressing the workload in the construction industry. A steady rather than widely fluctuating workload is desirable and important if the industry is to be allowed to manage its resources properly. The stop–go effect is more damaging than lower workloads. The adverse effect of recession in the industry is:

- unemployment of managers and operatives
- smaller firms being forced out of business
- larger firms reluctance to invest in new plant, equipment or other technology
- suppliers of materials and components being unlikely to extend their plants
- recruitment at all levels into the industry being made more difficult
- lack of continuity increases costs and decreases efficiency

The business cycle

The construction industry market, like the remainder of British industry, tends to develop a pattern known as the business cycle. Whilst alternating periods of booms and slumps in the construction industry coincide with movements of trade generally, they often occur earlier and last longer. As such they are often much more severe than experienced in other industries. In the nineteenth century, booms or peaks in activity tended to appear every 7–10 years. At these times construction industry activity was stretched in terms of both manpower and the availability of materials. In the early part of the 1960s, for example, a provisional order for materials such as facing bricks needed to be placed during the design phase in order to be sure of obtaining the materials at the appropriate stage during construction. Immediately after the end of the First World War the pattern altered and the country faced almost two decades of high unemployment with the slump reaching its trough in the depression of the mid-1930s. From the end of the Second World War until the early 1970s Britain and other world economies were running at relatively high levels, with little unemployment other than a blip in the late 1960s. Unemployment was never higher than 2.5% and was often as low as 1%. Concurrently with the stable employment levels the economy experienced a much lower inflation rate than the country was to face in the 1970s. This decade saw a return of the slump–boom pattern, but characterized by ever-deepening troughs and slumps and more modest recovery in the boom years. The time span had also fallen from a previous 7–10 year period to one of only 3–4 years.

The early 1970s were recession years followed by a 'bubble' in 1974 that was to burst by 1976. By 1978 the economy had started to pick up slightly but this was followed by a recession with over three million unemployed, many of whom should have been at work in the construction industry. This slump was worldwide with few countries escaping the human misery of high inflation. Since 1985 the economy rose to peak in 1989, but by 1990 was already in decline, and still in decline in 1992/93 causing unemployment to rise to even greater heights.

The construction industry is a useful indicator for economic commentators to follow since, historically, a decline or increase in construction activity has been a sign that recession or prosperity is to follow (Figure 1.1).

The multiplier effect

The general level of employment is dependent on an overall demand for goods and services in the country as a whole. If there is a high demand then more jobs are available as producers seek to increase their level of output. Conversely, if producers fear a fall in (world) demand for their goods, products or services, they invest less and in so doing create unemployment. The fall in employment will reduce demand and this results in an ever-decreasing spiral of loss in employment resulting in less demand for goods and services. This continues until conditions and, the key business criterion, confidence, cause some producers to expand and create the opposite effect to that described. This needs to be done carefully and cautiously otherwise unstable economic conditions can be created.

The construction industry as an economic regulator

Government is a major client of the construction industry and it is therefore tempting to suggest that the industry may be used as a regulator to control the economy. Whilst the industry is damaged by the stop–go nature of its activities, there is only scant evidence that government effectively turns the tap on or off in order to regulate economic performance. It may defer or cancel construction projects for other reasons, such as to reduce the public sector borrowing requirement which in turn may create a knock-on effect. Cuts in public expenditure sometimes have a high construction consequence, but these are often accompanied by other measures, so it is debatable whether this can be cited as an example of regulation.

However, government is able to intervene in the construction market in three ways; through finance, legislation or regulation and provision. The state can intervene in the market through finance by grants, benefits, subsidies and taxation. It can choose to offer grants for the construction of industrial or commercial premises in order to entice a potential factory owner to an area of high unemployment. It can also offer incentives for the construction of certain types of project such as private housing. It could also allow taxation relief against profits for the annual maintenance of building projects. There are all sorts of combinations of financial measures which can be used to stimulate construction activity if this is so desired. Changes in regulations, such as in town and country planning, can also create opportunities for construction development. For example, the allowance of a wider range of projects to be constructed in certain designated areas, may be just the sort of stimulus a developer is looking for. Thirdly, since government is a major client it has considerable scope to influence construction activity through the development, repair or maintenance of projects.

Housing

The opening caption Figure 1.1 (p. 1) in this book is, 'When new housing begins to be constructed the economy is moving out of recession.' Housing has been and continues to be a good indicator to not just of the measured prosperity of a country but also of its 'feel-good' factor. Throughout the early 1990s a majority of economic indicators have presented a positive picture of the economy, signalling the end of the recession:

• Inflation fell from over 20% at the start of the 1980s to between 2% and 3% by 1996.

- Bank base rates followed a similar pattern from over 15% to less than 5%.
- The current account deficit reduced from over £22bn to less than £5bn.
- Economic growth is expected to increase by 2–3%.
- Unemployment fell from a peak of 12% in the mid-1980s to less than 6% by the start of the 1990s.
- Whilst unemployment increased again at the start of the recession to over 10% but has reduced in almost every month since 1993 down to just over 2m.

However, against this background, houses, in sufficient quantities, were neither built nor sold due to the lacking ingredient of confidence in the future.

During the 1990s the housing market has been turned upside down. In the mid-1980s people were eager to buy. Gazumping was a common phenomenon and local authorities were encouraged by central government to sell their own housing stocks to any tenants who wanted the right to buy. Many house purchasers in the 1980s typically obtained mortgages that were up to four times their annual incomes.

Job insecurity has been a single and powerful influence on the housing market in recent years. Government figures put redundancy rates since 1990, at between 10% and 20% of the workforce. At the very least people have seen friends or colleagues lose their jobs at some point during those years. In many sectors of employment, people now no longer expect job security. Instead they envisage a working life with patches of unemployment, retraining and even a variety of different occupations. The mobility they need is enhanced through renting property rather than through its purchase. Whilst most people who are made redundant get another job within three to five months, the general ethos of insecurity has made people wary of committing themselves to the long-term obligation of mortgage interest payments.

Interest rates are currently very cheap encouraging people to purchase mortgages, but the memory of the high rates in the early 1980s and the inescapable possibility of higher rates at some time in the future makes people unwilling to take on a mortgage commitment.

House purchase has often been seen as an hedge against inflation and has in the past reaped considerable financial equity for many house owners. This scenario is not expected to occur within the current or immediate future economic climate, particularly with inflation being just a few percentage points.

Another factor mitigating against house purchase is the decreasing likelihood of state financial support to help with mortgage interest payments either through income tax relief, that declined in real terms in the 1980s, or in the event of a job loss.

There is also the risk of negative equity. Just before the last recession, with its wave of redundancies that severely affected the middle class home buyers, house prices rose to an all time high. House prices in 1989 were the highest they had ever been, even allowing for inflation. The reluctant readjustment in house prices was painful for those who were the last to jump on the housing bandwagon. Even by 1996, it was estimated that 40,000 mortgagees were still in negative equity home owning trap.

As the supply of private rented property failed to keep up with the booming demand caused by the above, private rents soared. This had been somewhat moribund for decades as a result of the 1950s' legislation giving security of tenure to tenants. Until 1988 there was no way that a UK landlord could be confident about access to rented property, or even the rent that could be charged. Landlords did not view rent assessments as fair or even viable. The 1988 Housing Act brought in Assured Shorthold Tenancies. These guaranteed the right of a landlord to repossess the property at the end of a tenancy and encouraged the provision of more private landlords.

With the real decrease in house prices and the reduction in the level of inflation, a house no longer seems to be a preferable asset to owning other commodities. Yet the income to price ratio has not been so favourable for many years. House purchase still makes sense, provided that the main motive of purchase is the instrumental one of providing a roof over one's head. It may be some time before this becomes a way of life for the majority again.

Indicators

There is a range of indicators which can be used to measure the performance of the construction industry. These can be used for an explanation of past activities and in the analysis of future trends. Some of the indicators are of a general nature which affect the performance across a wide range of industries whilst others are more specific measures of the construction industry alone. Table 1.11 presents the key economic indicators for the country as a whole. The fixed investment ratios represent a broader definition than the construction industry alone and include plant and vehicles. Nevertheless the indication from these figures of a boom being experienced throughout the country in 1988 was also present within the construction industry. The rapid decline in the early 1990s and the forecast for 1992 provided little cheer for the industry. This coupled with

Table 1.11   Key UK economic indicators [7, 9]

	1987	1988	1989	% change on year 1990	1991	1992	1993	1994	1995
Non-oil GDP	5.4	5.5	3.4	0.9	−2.0	2.0	2.0	3.5	2.4
Government consumption	1.2	0.4	0.9	2.8	2.3	1.5	−0.1	1.7	1.0
Consumers' expenditure	5.4	6.9	3.5	1.0	−0.9	1.9	2.5	2.6	2.0
Fixed investment	8.8	13.1	6.8	−2.4	−11.5	1.1	0.6	2.9	−0.1
Average earnings	7.8	8.7	9.1	9.7	8.1	6.3	3.8	4.4	2.8
Retail prices	4.1	4.9	7.8	9.5	5.8	3.6	1.9	2.9	3.2
Real disposable incomes	3.6	4.9	5.4	3.3	0.7	1.0	1.8	1.2	3.0
Three months interest rates [%]	9.7	10.3	13.9	14.8	11.5	9.5	5.4	6.6	6.5

high interest rates and a decline in disposable incomes make for gloomy forecasts [8].

Table 1.12 represents indicators which are of direct importance to the construction industry. These are based upon previous trends and future forecasts. They show that the construction boom of the late 1980s was coming to an end and that difficult times were faced by the typical contractor in the early 1990s. The decline in house building is partially due to the recession coinciding with a peak in a housing boom. The boom was sent into sharp reverse, house prices slumped and the collateral held by some mortgage lenders was insufficient to meet mortgage debts. Commercial building continued to expand but this was halted as the reported oversupply of office space in central London alone had reached 3 million square metres. The situation outside London was not much better with huge areas of freshly painted B1 office space being left empty. Many existing office blocks also had 'To let' signs indicating further empty space. In 1974 the private commercial construction output fell by £500m from £3.2bn down to £2.7bn. In 1990, the peak of the commercial output was £7.6bn falling to £5bn in 1992 and predicted to fall to £3.5bn in the current recession. Industrial construction was, for a time, still living off yesterday's corporate investment plans, but this fell sharply in 1990. Whilst the government began to spend again on public building, the long lead-in times of these schemes were unlikely to be felt until the beginning of 1994 [10].

Other indicators which are measured and recorded include, for example; the change in the number and volume of enquiries for work received by contractors, architects' workloads measured by the number of new commissions and projects that are at the production drawing stage, quantity surveyors workloads and materials production and deliveries. Information such as this is frequently recorded in the professional journals and forms a regular feature in *Building* magazine. A number of different organizations including the professional bodies, NEDO, DoE, BEC, material suppliers, product manufacturers and business information firms regularly analyse data and forecast predictions for the construction industry.

Trends

Future projected trends are measured on the basis of the previous performance of similar recorded data. Past trends may not necessarily be repeated. Table 1.14 represents time series data for some key indicators in the construction industry which measure the level of activity between the years 1965 and 1990. It can be observed that the late 1960s

Table 1.12 Fixed investment forecasts [7, 9]

	1989	1990	1991	% change on year 1992	1993	1994	1995	1996
By sector								
Private dwellings	-8.0	-17.1	-10.0	0.5	9.5	7.0	-8.4	-2.5
General government	24.7	10.5	-6.0	66.0	17.0	12.0	-7.3	-11.0
Business	7.9	-3.4	-13.0	-9.0	-8.5	-6.3	-0.1	0.7
By asset								
Dwellings	-3.7	-14.5	-9.0	9.5	11.0	8.0	-8.0	-4.6
Other buildings	11.8	8.0	-10.0	-2.0	-1.5	2.5	-2.5	-1.1
Total fixed investment	6.8	-2.4	-11.5	-4.0	-2.0	3.2	-0.9	0.2

Table 1.13 Construction output [7, 9]

					% change on year					
	1987	1988	1989	1990	1991	1992	1993	1994	1995	1996
Housing										
• starts			-21.5	-19.5	2.5	8.0	18.0	7.8	-14.7	1.0
• completions			-10.4	-8.8	-9.2	4.8	3.7	2.9	2.6	-5.5
Non-residential output										
• public	-5.2	1.0	5.3	11.0	1.5	-3.0	2.5	4.5	-2.8	-7.0
• industrial*	14.9	8.1	7.2	2.2	0.0	-5.0	-3.5	10.5	7.8	2.0
• commercial	18.1	14.6	26.4	11.3	-15.0	-20.0	-17.5	6.0	-1.4	6.0
Total	2.4	8.8	15.1	9.0	-7.0	-11.2	-8.5	6.3	-0.1	0.7
All new work	9.3	8.4	4.5	1.4	-6.4	-7.4	-1.5	2.5	-2.9	-1.1
Repair, maintenance & improvement	6.0	4.4	3.2	-0.3	-10.6	3.5	-2.5	4.5	2.0	2.0
Total all work	7.8	6.6	4.3	1.0	-9.0	-4.0	-2.0	3.2	-0.9	0.2

*Estimated to exclude the Channel Tunnel project, an allowance for which is included in the total work

Table 1.14   Indices of construction trends [4]

	New orders	Output of work	House building Starts	House building Completions	Brick deliveries	Cement deliveries
1965	145	104	196	196	192	127
1966	136	106	190	197	173	127
1967	156	112	224	207	198	132
1968	144	115	197	212	186	135
1969	134	114	172	188	167	131
1970	132	111	160	179	164	128
1971	141	113	172	179	176	133
1972	149	115	176	163	181	135
1973	145	116	164	151	180	130
1974	101	104	126	138	129	132
1975	100	98	162	160	141	127
1976	106	96	163	161	122	110
1977	95	96	134	155	–	–
1978	100	103	132	143	131	112
1979	91	105	113	125	126	115
1980	75	100	78	120	104	106
1981	81	90	77	102	92	93
1982	86	91	97	90	97	96
1983	99	96	111	102	107	99
1984	103	99	99	107	105	101
1985	100	100	100	100	100	100
1986	108	103	107	104	108	101
1987	129	111	115	109	113	107
1988	134	119	126	116	121	126
1989	130	125	100	105	102	127
1990	111	126	81	95	89	–
1991	104	117	66	89	80	104
1992	101	113	63	83	75	95
1993	113	111	75	85	81	95
1994	111	115	81	87	90	95

(1985 = 100)

represented good years for the construction industry, 1967 being the best year over this period for new orders, housing starts and brick deliveries, and 1968 achieving the most housing completions and one of the best years for cement deliveries. Conversely, the early 1980s represented poor years for the industry with low values across most

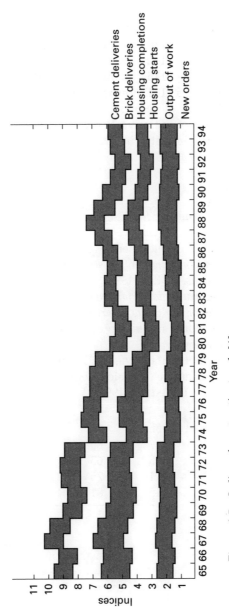

Figure 1.5 *Indices of construction trends* [4]

of these indicators. New orders were particularly poor between 1979 and 1983 and this had a knock-on effect on the output of work between 1981 and 1984. Housing starts from 1980 onwards were at about half the level of the late 1960s and evidence suggests that the low number of starts registered in 1990 will continue throughout the early part of the decade. The NEDO forecasts [11], indicated that the 161 000 housing starts in 1990, which were the lowest since 1981 (155 000) were unlikely in the short term to reach many more than 190 000 units. Throughout the 1980s an average of 200 000 units per year were started, with over 250 000 housing starts in 1988. The government's gradual withdrawal of public sector work and the vagaries of the peaks and troughs of the private sector require governments to act as a moderator if the industry in the future is to become efficient in its practices.

References

[1] *Lloyds Bank Economic Profile of Britain 1996.* Lloyds Bank
[2] *Department of Employment Labour market and skill trends 1996*
[3] Department of Employment. *Annual abstract of statistics.* HMSO
[4] *Housing and Construction Statistics.* HMSO, 1995
[5] *Regional Trends.* HMSO, 1991
[6] *Britain 1996, An Official Handbook.* HMSO, 1996
[7] *Barclays Economic Review Quarterly.* Barclays Bank
[8] 'Riding the storm' in *Building,* 11 January 1992
[9] *Building* (various)
[10] 'Hitting the bottom' in *Building,* 10 January 1992
[11] National Economic Development Office. *NEDO Construction Forecasts 1991–1992–1993.* 1991

2
Development

The challenge for the modern architect is the same as the challenge for all of our lives: to make out of the ordinary something out-of-the-ordinary.

Patrick Nuttgens, 1979

Figure 2.1   *Development – the creation of the construction project*

Introduction

Development in construction involves, first of all, identifying a need for a project. This need may be to satisfy the placing of investment funds or to address the requirements in society for housing, health or education together with the relevant infrastructure. Factors such as population trends, changes in patterns of lifestyles and fashion will all need to be considered. The selection of an appropriate site, the choice of consultants and funding sources are the three early decisions for the client or promoter to make. Development is the creation of construction (Figure 2.1).

The development cycle

The development cycle of a construction project can be separated into five stages as shown in Table 2.1. These are not discrete functions and an overlap of the different activities occurs at the various stages of the

Table 2.1   The development cycle

Stage	Phase	Typical time duration (years)
Inception	Brief	1
	Feasibility	
	Viability	
Design	Outline proposals	
	Sketch design	
	Detail design	1
	Contract documentation	
	Procurement	
Construction	Project planning	
	Installation	3
	Commissioning	
In-use	Maintenance	
	Repair	80
	Modification	
Demolition	Replacement	–

project's life cycle. Emphasis should be on securing developments that best satisfy all of these objectives rather than relying upon an appraisal of the initial expectations alone.

The inception stage

This is the stage where the client will be determining the objectives for the project. A majority of projects arises from a long planning programme where the clients or promoters are considering the scheme as a part of the overall objectives of their own organization. For example, in terms of the schools' capital building programme there are detailed plans kept by the Department for Education and Employment (DFEE). These are based upon an appraisal of the existing building stock, the changes in the age ranges of the school children, regional population trends and the requirements from local education authorities for existing school replacement or renovation. This information is then set against the availability of capital funding which is agreed in the government's annual expenditure forecasts and budgets. The many different scheme proposals up for consideration always exceed the amount of funds which are allocated to this type of work. Some of the funding must also be set aside for emergency works which are likely to occur each year to rectify vandalism, fire damage and other emergency repair work.

The clients who are involved in one-off projects often come to their chosen consultant with a broad outline of their aspirations, a sum of money which is often insufficient and a time scale for occupation which is often impossible. The better informed clients, who are those involved in frequent capital development, usually have more realistic expectations of what can and cannot be achieved. The type of project will often determine whom the client or promoter appoints as designer. On engineering projects, civil engineering consultants are the most likely choice. For building projects generally the architect has traditionally been the first point of contact with the client. These are traditions which die hard. However, on smaller works and schemes of refurbishment the building surveyor is being increasingly used as the client's main advisor. As the different combinations of design and build or management contracting are employed, clients now often appoint the construction firm direct, choose an alternative consultant as a main partner in the venture or appoint a project manager in overall charge of the scheme. Table 2.2 indicates the typical time span of activities for different sorts of projects. The long lead-in times at inception for some projects reflect the need to obtain the required finance and planning approvals.

Table 2.2   Characteristic times involved in the various stages of the development cycle [1]

	Inception stage (years)	Design stage (years)	Construction stage (years)
Public sector			
Housing	1–4	1–3	1–4
Health	1–5	0.5–4	0.5–5
Education	1–4	0.5–3	0.5–2.5
Other large buildings (lawcourts, civic buildings, etc.)	1–7	1–3	1.5–2.5
Other small buildings (libraries, etc.)	0.5–3	0.5–2	0.5–1.5
Roads & harbours	1.5–10	1–4	0.5–3
Water & sewerage	1–4	0.5–3	0.5–1.5
Private sector			
Housing	0.5–6	0.5–4	0.5–1.5
Industrial	0.5–2	0.5–2.5	0.5–2
Commercial	1–10	1–4	0.5–3

Client's requirements

It is necessary in the first instance to determine the client's main requirements. Clients of the construction industry are wide and diverse, each with their own particular needs and desires regarding their project. The wide variety of contractual procedures which are now available reflect this fact. Principally the client's requirements can be attributed to the following four main factors:

Time
• length of time required for the design
• start and length of the contract period
• completion by the date stated in the contract
• time certainty

Cost
• initial cost and relationship to tender sums and final cost
• value for money
• whole life cycle cost approach

Performance
- design in terms of function and appearance
- construction reliability and performance
- no latent defects and trouble free guarantees
- low time and cost maintenance

Management
- clear allocation of responsibilities
- accountability; particularly in the public sector
- clear evaluation of risks

Table 2.3 offers a numerical analysis of the client's requirements. From the information provided in the client's brief, it should be possible to translate the ideas into tangible plans and designs, which at least provide an initial framework for the project appraisal.

Table 2.3   Client's requirements [2]

Quality	45	Technology	15
		Function	25
		Aesthetics	5
Costs	35	Initial	20
		Life cycle	15
Time	20		
Client total	100		

The feasibility phase seeks to determine whether the project is capable of execution in terms of its physical complexities, planning requirements and economics. The available site, for example, may be too prohibitive in terms of its size or shape or the ground conditions may make the proposed structure too costly. Planning authorities may refuse permission for the specific type of project or impose restrictions that limit its overall viability, in terms of, for example, the return on the capital invested. Schemes may be feasible but may not be viable. Office development in the City of London, which remained empty in 1991, was shown not to be a viable proposition at the time of its completion, due to the general over-supply of this kind of accommodation. Debenham, Tewson and Chinocks [3], a firm of property managers, published a report in 1991 indicating that there were available over 3 million square metres of empty office accommodation in central London alone. In the City itself there had been an increase in vacancy rates by over 20 000 square metres to 1.5 million square metres. This represented 18% of the City offices accommodation.

The design stage

The alternative options and advantages of choosing a designer separate from the contractor, or to use design and build in preference, are well documented. In any event it is first necessary to prepare schematic outline proposals for approval, prior to a detailed design. These proposals will need to be accepted by the client in terms of the requirements which have been outlined in the brief; by the relevant planning authorities, for their permission; and in terms of an outline budget by way of an initial cost plan. As the scheme evolves and receives its various approvals, a number of different specialist consultants will be employed. Some of these may be public relations consultants, particularly where a sensitive scheme such as a new road, building in the green belt or where a project which is out of character with the locality is being proposed.

The detailed design will follow where all of the previous activities have been agreed and approved by the client. Different solutions to spatial and other design problems will be considered and some of these are likely to have a knock-on effect on aspects of the project which have already been agreed. Each alternative solution will need to be costed to ensure that the cost plan remains on target and, where they significantly affect the client's proposals, they will need to be communicated to the client for agreement. It will probably be necessary during this stage to involve firms who supply and install the specialist equipment which is required. The documentation which is required for tendering purposes will also be prepared at this time and concurrently as the design develops. When the project is approaching the tender stage, the different firms which may be interested in constructing the project should be invited to tender. Upon the receipt of the documentation the contractors enter their estimating phase, since the awarding of the works of construction is most frequently done through some form of price competition. Table 2.2 suggests design periods which may be required for some typical projects. The long periods of time reflect the complexity which is required for solutions to one-off designs for construction projects.

The construction stage

This stage commences when the contractor begins the work on site. It is often referred to as the post-contract period, since it commences once the contract for the construction of the project has been signed and work has started on site. Where the project is on a design-and-build arrangement or a system of fast-track procurement then this stage may start before the design is finalized and then run concurrently with it. Contractors are critical of the traditional arrangements since they are frequently required to price the works which, although assumed to be fully designed, are in reality not so. Throughout this stage formal instruction orders are given to the contractor for changes in the design and valuations are prepared and agreed for interim payment certificates. Contractual disputes unfortunately arise all too often due to misunderstandings or incorrect information being made available. The contractor is also sometimes over-ambitious and enters into legal agreements that become impossible to fulfil. These create grounds for damages on the part of the client. Project completion times can last from anything from a few months up to 10 years or more. Upon completion, the formal signing over of the project to the responsibility of the client is made and the project enters the third stage in its life cycle.

The in-use phase

This is the longest stage of the project. The immediate aims of the client are now hopefully satisfied and the project can be used for the purpose of its design. During this stage occasions will occur when maintenance will be required. Even so-called 'maintenance-free construction' requires some sort of attention! The correct design, selection of materials, proper methods of construction and the correct use of components will help to reduce maintenance problems. A sound understanding, based upon feedback from project appraisals in practice, will help to reduce the possible defects that can arise in the future. Defects are often costly and inconvenient and minor problems sometimes require a large amount of remedial work to rectify; sometimes out of all proportion to the actual problem. Many projects have only a limited life expectancy before some form of refurbishment or modernization becomes necessary. The introduction of new technologies also makes previously worthwhile components obsolete. City centre retail outlets have a relatively short life expectancy before some form of extensive refitting becomes required. Fifteen years seems to be an optimum age. Whilst the shell of buildings may have a relatively long life of up to 100 years, and some are able to last for centuries, their respective components wear out and need frequent replacement as suggested in Table 2.4. Obsolescence is also a factor to consider in respect of component replacement.

Table 2.4   Component lives in buildings

	Expected life span (years)
Bathroom-fittings	20–25
Kitchen-fittings	10–15
Windows & doors – softwood	10–15
– hardwood	20–25
Central heating systems	30–35
Central heating boilers	10–15
Brickwork pointing	20–25
Electrical installations	25–30
Interior decoration	3–10
Exterior decoration	2–5
Pitched roofs	40–50
Flat roofs	8–12

Demolition

The final stage in a project's life is its eventual demolition, disposal and the possible recommencement of the life cycle on the same site. Demolition becomes necessary through decay and obsolescence and when no further use can be made of the project. Some buildings are destroyed by fire, vandalism and explosion, and may become danger-ous structures that require demolition as the only sensible course of events. There are relatively few projects that last forever, and become historic monuments, since the style of life and the needs of space is constantly evolving to meet new challenges. Some notable projects become listed buildings. The Secretary of State for the Environment has powers under the planning acts to compile lists of buildings which are of special historic interest. It then becomes an offence to demolish, alter or extend a listed building in any way that would affect its character as a building under this regulation. Where non-listed buildings are thought to have special historic or architectural interest then a planning authority may serve a building preservation notice upon the owner. Demolition can be achieved in different ways, from 'hand' demolition, where parts of the structure are removed a piece at a time, to large structures that can apparently be reduced to a heap of rubble in a few minutes.

Architects' workloads

Table 2.5 represents architects' workload figures, showing the percentage change occuring over a 2-year cycle. These provide a good indicator of the future fortunes of the contractor and the industry, particularly where new commissions are translated into working drawings. However, even at this stage contractors cannot be sure that such schemes will actually start on site, and even then there are examples of projects which have been temporarily abandoned in the hope that they will be restarted when better times arrive. The figures do not attempt to measure changes in the way that commissions were received, i.e. from clients or through design and build contractors who choose to employ a private firm of architects. The architects' workload provides a sorry picture of their activities at the start of the 1990s. Similar sorts of figures are recorded by the other professions on the workloads of their members, e.g. quantity surveyors' workloads by the RICS.

Property developers

There are about 100 quoted property companies listed in the FT-SE index. Table 2.6 lists the top seven of these companies in terms of

Table 2.5   Architects' workloads [4]

	New commissions (Percentage change)	Production drawings	Contractors output
1970	+15	+5	0
1972	+35	+35	+5
1974	−10	−10	−10
1976	−5	−5	0
1978	+30	+40	+5
1980	−20	−10	−5
1982	+10	−5	0
1984	+10	+15	+5
1986	+30	+25	+5
1988	+25	+10	+5
1990	−25	−15	−5
1992	-40	−30	−15
1994	−5	−5	+5
1996	+30	+5	+5

Table 2.6   The major property companies listed in the *Financial Times*, August 1996 [5]

	Share price	Market capitalization	Yield	Price/Earnings Ratio
HK Land	134	3634	5.5	15.1
Land Securities	679	3275	5.1	18.9
Land Lease A	950	2249	3.7	30.0
British Land	430	1853	2.5	36.2
MEPC	415	1738	6.0	21.4
Hammerson	370	1049	3.6	24.2
Capital Shop Centres	286	1041	3.3	28.2

their market capitalization values. All of the property companies saw their capital values and profits decline at the beginning of the 1990s. Several went into receivership or merged with other property companies that were experiencing similar difficulties. Many had accumulated large debts. The difficulty with property development is the long lead-in time and, like a super oil-tanker, once on the move developments are difficult to curtail or abandon when the direction of the market changes. This happened with the Canary Wharf development that was begun in the 1980s at a time of huge demand for office space in and around London. The financial collapse occurred because events, even just a few years ahead of the 1980s boom, could not be anticipated let alone forecasted. Since the depression of the early 1990s, property market values are now beginning to recover to more sensible levels.

A recession, a fall in stock market prices can trigger a fall in capital values. These will also reduce when property prices generally begin to fall, as occurred in the late 1980s and early 1990s. The decline in the property asset values of industrial companies and financial institutions can also lead to a decline in investment and lending. A report, published in 1991, based upon research carried out by the London Business School [6] made the following observations:

- Commercial property was worth £250bn and is essential to the activities of manufacturing, services and agriculture.
- The total value of commercial property in 1989 was more than double that of outstanding government gilts, and about one half of the value of the total UK equity market.

- Economic activity directly concerned with commercial property and construction contributed just under 6% of total GDP. A broader contribution of the sector, results from its role as an essential factor of production. This is estimated to be around 10% of GDP.
- The direct contribution of commercial property and construction to GDP is the same size as the UK energy sector, about a quarter of the contribution of the manufacturing sector and about one-third of the banking and financial sector.
- Commercial property accounts for one-third of total investment in physical assets in the economy which is about the same as investment in plant and machinery.
- Property represents an important asset in the balance sheets of companies, representing an average of 150% of net assets, 30% to 40% of total assets and 100% of capital.
- Commercial property accounts for about 20% of the total assets of financial institutions.
- Banks and institutional lending to industrial companies is based partly on profit projections but partly on collateral provided by the assets of the firm.

Delays in development

In 1991, according to Applied Property Research [7], 40 of the 75 major office schemes planned for London worth an estimated £11 157m had yet to get the go ahead by the developers. These projects represented 1.5m square metres of floor space, had already received planning permission and in some cases a contractor had already been appointed. Figures also showed that the designs for 20 of the schemes, which represented half a million square metres of space and worth £430m had gained planning permission two years earlier. The leading property developers have stated that due to these delays there will be no new speculative office building available until at least 1997, even if work began today. Much of it may also need to be extensively redesigned due to changes in requirements and fashion that have occurred during this period of time. The RIBA reported that during 1991/92 on average 40% of commissions were abandoned. The figure was over 80% in the north of England. Vacancy rates in city centre offices in mid 1991 were running at 20%, which is equivalent to the entire office stock in Birmingham and Manchester. Rents fell by about 40% between 1988 and 1991. The banks, which account for a large proportion of the lending, want to reduce the debt which is owed by property companies. This will impact further upon future developments. According to another report [8], the cause of the property crash in the 1990s was due to: over-ambitious developers (36%), high interest rates (30%), poor lending (20%), overinflated valuations (8%) and a lack of institutional investors (6%).

New development

During the latter half of 1996 the beginning of the end of the recession in the construction industry became a possibility. With the glimmer of rising rents for commercial property, new development has again become feasible. The property consultants, Richard Ellis, has reported that 1.6m square feet (160 000 square metres) of offices will be constructed in London alone during 1996–98 by a number of property developers. A key test of confidence will be the release in the early part of 1997 of speculative offices by a number of firms such as British Land, Equitable Life and City of London Properties. Much of the speculative development has been driven by the developers and financial institutions taking the view that the depressed property market can only improve.

Rents have increased to £35 per square foot (£350 per square metre) and in some cases to as much as £40 per square foot (£400 per square metre). Such increases in rents will be slowed down if the Canary Wharf development embarks upon a new phase under its new owners. There is a constant rivalry between Docklands and the City of London. This factor was one of the reasons for the demise of Olympia and York, the developers of Canary Wharf. Some of the large financial institutions have considered moving to Canary Wharf because of the shortage of suitable premises in the City. Some analysts are arguing that even on conservative projections the current supply of office space will be insufficient to meet tenant demand.

Office rentals

Table 2.7 provides an example of office rents charged in different parts of the world. This indicates that London is one of the most expensive cities of Europe for the rental of office accommodation. However, on a world-wide basis, cities in the far east are shown to be a more expensive location.

In the private sector office accommodation supposedly benefits from a 'prestige' address, where firms have required their offices in the larger cities and in the expensive inner urban areas of those cities, London being a prime example. By comparison the respective costs in the provincial cities in Britain were much lower. For example, in some

Table 2.7   Typical World Office Rental levels (1996) (Source: *Property Week*)

Location	Rent £/m²	Service charges %	Occupation charges* £/m²
Bombay	1076	9	1168
Tokyo	775	19	915
Hong Kong	769	9	872
Moscow	603	14	683
Beijing	565	6	608
London (West End)	457	15	758
Paris	419	14	516
Geneva	393	7	414
London (City)	377	19	710
Frankfurt	334	14	387
Berlin	264	17	312
Milan	199	23	247
Amsterdam	182	16	210
Johannesburg	59	23	75

* excluding rates

northern cities, prime city centre rates were less than 10% of the London rates. However, even taking these wide disparities into account, typically 70% of all new office investment has been located within the London area. When occupancy costs (rent, rates, service charges) are taken into account the London rates reached as high as £650 per square metre.

Traditionally offices have tended to be located in the central business districts because of:

• the interrelationship of businesses requiring ready access to other offices
• dispersion is uneconomic
• convenience of access to their staff and the public in respect of public transport systems
• access to local and central government offices

Such high costs often mean that much of the central areas are given over to offices with perhaps the ground floors being used for retailing, banking, etc. where the public needs to gain easy access. The earning capacity of some offices is sometimes dependent upon being sited in a specific area, where face-to-face contact can be made and access to the required information and data is easily retrieved. Since these sites are the most valuable it becomes practicable to construct multi-storey structures to gain the maximum benefit. (For additional information see *Precontract Studies* by Allan Ashworth (Addison Wesley Longman 1996) [9]).

Investment

Financial investment of any kind relies upon the analysis of three variables: risk, return and time. A good investment can therefore be described as one which produces the rate of return expected within the time required having regard to the risk involved. The considerations to be weighed in property investment are as varied as the number of funds seeking to invest. A fund manager's strategy will want to consider: freehold or leasehold, geographical spread, type, e.g. shops, offices, industrials, etc., pattern of rent reviews, size and yield, age and quality of buildings, type of covenant and tenant. The perfect balance will rarely be achieved. The main factor to consider with any property is location. For example, a dilapidated building sandwiched in the high street between Marks & Spencer and Mothercare is better than a modern shop at the wrong end of town. Most of the investment is also likely to be in the building site rather than the building. The second factor is obsolescence. Early 1960s offices, which are less suitable for modern open layouts or the installation of communications, or warehouses with an eaves height too low for the current generation of fork-lift truck, are examples of buildings that have become outdated within a very short space of time. The third factor is that of marketability. Investors dislike being locked into a property which will be difficult to value or to re-let.

The banks lending to property companies (Table 2.8) increased sharply in the latter half of the 1980s, and in 1992 stood at 8.5% of total bank lending [10]. This is almost at the level that triggered the secondary banking crisis in the mid-1970s. However, the Bank of England is keen to stress the difference between the 1974/75 crisis and that of the early 1990s in two respects; the heavy exposure to property of the weakly capitalized secondary banks and the limited role of the foreign banks in the British property market. In the 1990s the leading British banks and foreign investment are much more involved. The total bank lending to British property companies is just over £39bn. Of this just over £17bn is owed to foreign banks, including £4.5bn to the Japanese. Whilst the City and Docklands are trou-

Table 2.8   Clearing bank lending to property companies [10]

| | Clearing banks UK property lending (£m) | | | | | |
	1986	1987	1988	1989	1990	1991
Barclays	2 000	2 100	3 100	4 400	5 200	5 400
Lloyds	1 500	1 700	2 500	3 900	4 400	4 100
NatWest	1 000	1 400	2 000	2 600	2 800	2 400
Midland	600	700	1 400	1 900	1 900	1 600
Totals	5 100	5 900	9 000	12 800	14 300	13 500

Note: total lending by all banks was worth about £39bn in 1991

bled areas, the banks' lending on property in other parts of the country is performing well. The direct risk to the banks is also limited since they have been operating tougher rules on the ratio of capital to outstanding loans.

Approvals and controls

Government has wide powers of control over the development of construction works. It seeks to resolve the conflicting demands of industry, commerce, housing, transport, agriculture and recreation by means of a comprehensive statutory system of land use planning and development control. The government's aim is for the maximum use of urban land for new developments, with the intention of protecting the countryside and assisting urban regeneration (Table 2.9).

The system of land use planning in Britain involves a centralized structure under the Secretary of State for the Environment. The

Table 2.9   Urban land composition

Housing	45–50%
Industry	5–15%
Offices	10–25%
Shopping	15–20%
Education	5–10%
Open spaces	15–25%

The area of agricultural land transferred to urban areas per year since 1924, is about 1500 hectares per annum.

strategic planning is primarily the responsibility of the county councils, while the district councils are responsible for local plans and development control, the main housing function and environmental health. The development plan system involves the structure and local plans. The structure plans are prepared by the county planning authorities and require ministerial approval. They set out the broad policies for land use and ways of improving the physical environment. Local plans provide detailed guidance, usually covering about a 10-year period. These are prepared by district planning authorities but must conform with the overall structure plan. In 1989 the government announced its intention to simplify the two-tier planning system in the non-metropolitan areas by replacing it with a single tier of district development plans.

Before a building can be constructed, application for planning permission must first have been obtained. This is made to the local authority in the form required by the Town and Country Planning (Making of Applications) Regulations. In the first instance outline approval would be sought to avoid the expense of a detailed design which could fail to secure approval. Full planning permission must still be obtained. If a scheme fails to obtain approval then a planning appeal can be made to the Secretary of State. The Building Act [11] was introduced onto the statute book in 1984. The 1985 Building Regulations are framed within this Act and allow two administrative systems to be applied; one through the local authority building control department and the other via certification.

Such controls are necessary to:

- secure improved standards of design and construction
- ensure the safety and health of the occupants
- provide for the proper location of buildings and industry
- make the best use of the land which is available
- provide for the safety, health and welfare of those engaged in the construction process and those affected by it

The contravention or disregard of the laws relating to building and construction, whether intentionally or unknowingly, will render the offender liable to prosecution and in some cases to imprisonment. Some of the more common Acts of Parliament are:

- Town and Country Planning Acts 1990
- Housing Act 1985
- Local Government Planning and Land Act 1988
- Highways Act 1980
- Environmental Protection Act 1990
- Derelict Land Act 1982

Facilities management

The development of all capital works projects results in buildings and structures that require adequate maintenance if they are to be used effectively and efficiently throughout their lives. Traditionally building maintenance management has been in the province of the building surveyor, although other professions, notably architects and engineers, have also had major roles.

During the 1990s, a new activity of facilities management (FM) began to grow rapidly from a relatively small base. FM is concerned with building repairs and maintenance but this is only a part of its activities. FM is concerned with managing not just the building structure but the facilities that are also provided within the structure in order to improve the overall value for money for the client.

The Centre for Facilities Management at Strathclyde University claims that the potential value of the FM in Great Britain is about £160 billion or about almost four times the size of the construction industry. A report published by the Centre shows that the providers of FM services are being given increased levels of responsibility and are now involved in budget setting, the development of service level agreements and liaison with customers. The survey also found that facilities managers generally report directly to the board of companies, have responsibility for large portfolios covering several sites and typically manage budgets of £2.5m. They prefer out-sourcing contracts to be short term and to cover a single service.

Going green

The concept of a green building is an elusive one [12]. The definition is broad and being green in a professional sense may merely come down to a change in attitude. Most buildings are designed to cope with the deficiencies of a light loose structure, designed to meet the Building Regulations thermal transmittance standards and no more. The fact that our Building Regulations are a long way behind those of other European and Scandinavian countries is not seen as a cause for concern. The fact that about 56% of the energy consumed, both nationally and internationally, is used in buildings, should provide designers with opportunities and responsibilities to reduce global energy demand. There is a need to make substantial savings in the way that energy is used in buildings, but there is also a need to pay attention to the energy used in manufacture and fixing in place of a building's components and materials. For a new building this can be as high as five times the amount of energy that the occupants will use in the first year.

In the 1950s and 1960s building maintenance and running costs were largely ignored at the design stage of new projects. Today the capital energy costs which are expended to produce the building materials and to transport them and fix them in place is ignored in our so-called energy efficient designs. In any given year the energy requirements to produce one year's supply of building materials is a small but significant proportion (5–6%) of total energy consumption, which is typically about 10% of all industry energy requirements. The building materials industry is relatively energy intensive, second only to iron and steel. It has been estimated that the energy used in the processing and manufacture of building materials accounts for about 70% of all the energy requirements for the construction of the building. Of the remaining 30%, about half is energy used on site and the other half is attributable to transportation and overheads. Although the energy assessment of building materials has still to be calculated and then weighted in proportion to their use in buildings, research undertaken in the USA has shown that 80 separate industries contribute most of the energy requirements of construction and five key materials account for over 50% of the total embodied energy of new buildings. This is very significant since considerable savings in the energy content of new buildings can be achieved by concentrating on reducing the energy content in a small number of key material producers.

References

[1] NEDO. *How flexible is construction. A study of resources and participants in the construction process.* 1978

[2] Walker, A. *Project Management in Construction.* Collins, 1984

[3] Debenham, Tewson and Chinocks. Central Offices Research. D, T&C, 1991

[4] Royal Institute of British Architects (various)

[5] Klienwort Benson Securities and *Chartered Surveyor Weekly*

[6] Kynoch, R. and Goodman, T. 'Will the crunch come as values fall?' in *Chartered Surveyor Weekly*, 31 January 1991

[7] Duffy, A. 'London office projects fail to get the go ahead' in *New Builder*, 7 May 1992

[8] Kynoch, R. 'Banks say bad lending is partly to blame for crash' in *Chartered Surveyor Weekly*, 7 March 1991

[9] Ashworth, A. *Precontract Studies.* Addison Wesley Longman, Harlow, 1996.

[10] Olins, R., Lynn, M. and Smith, D. 'Disaster in Docklands' in *Sunday Times*, 31 May 1992
[11] The Building Act 1984
[12] 'Going Green' in *Building Services Engineering*, October 1990.

3

The construction product

Studies to improve quality are essential.

Building Towards 2001. *Building*, 1990

The product

The official statistics which specify the value of the construction output in Great Britain are grouped in accordance with the Standard Industrial Classification (Revised 1980). This covers:

General construction and demolition work
Establishments engaged in building and civil engineering work, not sufficiently specialized to be classified elsewhere in Division 5, and demolition work. Direct labour establishments of local authorities and government departments are included.

Construction and repair of buildings
Establishments engaged in the construction, improvement and repair of both residential and non-residential buildings, including specialists engaged in sections of construction and repair work such as bricklaying, building maintenance and restoration, carpentry, roofing, scaffolding and the erection of steel and concrete structures for buildings.

Civil engineering
Construction of roads, car parks, railways, airport runways, bridges and tunnels. Hydraulic engineering, e.g. dams, reservoirs, harbours, rivers and canals. Irrigation and land drainage systems. Laying of pipe-lines, sewers, gas and water mains and electricity cables. Construction of overhead lines, line supports and aerial towers.

Construction work at oil refineries, steelworks, electricity and gas installations and other large sites. Shaft drilling and mine sinking. Laying out of parks and sports grounds. Contractors responsible for the design, construction and commissioning of complete plants are classified to heading 3246. Manufacture of construction steelwork is classified to heading 3204. The treatment of installation work is described in the introduction to the SIC.

Installation of fixtures and fittings
Establishments engaged in the installation of fixtures and fittings, including such as gas fittings, plumbing, heating and ventilation plant, sound and heat insulation, electrical fixtures and fittings.

Building completion work
Establishments specializing in building completion work such as painting and decorating, glazing, plastering, tiling, on-site joinery and carpentry, flooring (including parquet floor laying), installation of fireplaces, etc. Builders' joinery and carpentry manufacture is classified to heading 4630; shop and office fitting to heading 4672.

Table 3.1 shows the construction output for 1970–1994 [1] in (i) actual terms (Figure 3.1), and (ii) real terms at 1985 prices (Figure 3.2) [2]. The construction output is the output by contractors, including unrecorded estimates by small firms and self-employed workers; it also includes that of the direct labour departments. Examination of Table 3.1 shows that, over these two decades, the total output in real terms has demonstrated cyclical variation with the minimum output some 60% of the maximum. Minimum outputs occurred during recessions in 1977 and 1981. The total output peaked in 1990 at £53 307m recovering from the depths of one of the worst post-war recessions in 1981 of £21 547m. In real prices this is a substantial increase in output of nearly 40%. Output again fell during the past years confirming the industry's reputation for cyclical behaviour. There has been a steady increase in output during the 1980s. All sectors showed an increase with the exception of housing in the public sector which has shown a steady decline.

In 1990, new buildings accounted for 56% of the construction industry output and repairs and maintenance 44%. Some 22% of the new work was in housing and 78% in other construction work. Of the new building, 23% was for the public sector and the remaining 77% for the private sector. Due to pressure to control public expenditure during the 1980s, the public sector's share has declined from 40% of new construction in 1980 to the latest figure of 24% in 1994. Both these

Table 3.1   Construction: value of output in Great Britain [1, 2] (all prices in £1000m)

	Year 1970	1971	1972	1973	1974	1975	1976	1977	1978	1979	1980
All work: total	5.36	5.93	6.75	8.61	9.73	11.08	12.18	13.31	15.70	18.87	22.05
All work: total (at 1985 prices)	30.99	31.36	31.96	32.27	28.96	27.27	26.84	26.71	28.75	29.18	27.83
New work: total	3.83	4.27	4.79	6.13	6.80	7.72	8.48	8.97	10.31	11.72	13.06
New housing: total	1.40	1.59	1.87	2.49	2.54	2.96	3.52	3.55	4.12	4.38	4.30
For public sector	0.68	0.67	0.68	0.85	1.10	1.45	1.76	1.72	1.75	1.71	1.71
For private sector	0.72	0.92	1.19	1.64	1.44	1.51	1.76	1.83	2.38	2.67	2.59
Other new work: total	2.43	2.69	2.92	3.64	4.27	4.76	4.96	5.43	6.19	7.34	8.76
Infrastructure: total											
For public sector	1.27	1.36	1.49	1.81	2.00	2.35	2.57	2.60	2.76	3.07	3.52
For private sector: total	1.16	1.33	1.43	1.83	2.27	2.41	2.39	2.82	3.43	4.27	5.24
Industrial	0.61	0.64	0.68	0.83	1.05	1.15	1.19	1.51	1.80	2.35	2.81
Commercial	0.55	0.69	0.75	1.00	1.22	1.26	1.20	1.31	1.62	1.92	2.43
Repair & maintenance: total	1.53	1.66	1.96	2.49	2.93	3.35	3.70	4.34	5.39	7.15	9.00
Housing: total	0.67	0.75	0.93	1.26	1.47	1.63	1.76	2.07	2.57	3.60	4.48
For public sector	0.59	0.64	0.74	0.86	0.97	1.21	1.34	1.53	1.85	2.27	2.92
For private sector	0.26	0.28	0.29	0.37	0.48	0.52	0.61	0.74	0.96	1.28	1.60
Public other work											
Private other work											

Note: Separate figures for public and private sectors repair and maintenance for housing unavailable before 1985.

figures contrast with 1970 when the public sector accounted for some 51% of new work. The reduction in public sector spending and greater dependency on the private sector has exposed construction more fully to market forces. Consequently, demands and cutbacks are not so easily moderated with the former accentuated and the latter deepened. This has been the pattern of recent years. It has been more difficult for the construction industry to respond effectively under these circumstances. During periods of recession tender prices fall significantly and the construction industry offers greater value for money. At these times there is a stronger case for increasing public expenditure which would have the added benefit of alleviating unemployment.

Returning to 1990, the biggest individual item of new work was commercial work for the private sector at £11 310m. This work

1981	1982	1983	1984	1985	1986	1987	1988	1989	1990	1991	1992	1993	1994
21.55	22.54	24.34	27.76	29.66	32.05	37.53	44.85	52.15	55.31	51.11	47.47	46.32	49.44
25.14	25.47	26.64	29.31	29.66	30.64	34.17	37.46	39.48	39.87	37.17	35.68	35.00	36.13
12.35	12.63	13.40	14.54	15.30	16.68	19.90	24.76	29.32	30.76	27.71	24.81	23.56	25.09
3.74	3.92	4.85	4.95	4.64	5.43	6.85	8.66	8.09	6.68	5.79	6.08	6.63	7.42
1.22	1.02	1.12	1.04	0.84	0.76	0.86	0.88	0.97	0.93	0.79	1.24	1.42	1.67
2.52	2.90	3.73	3.91	3.80	4.67	5.99	7.78	7.12	5.75	5.00	4.84	5.21	5.75
8.62	8.71	8.55	9.69	10.66	11.25	13.05	16.10	21.23	24.08	21.92	18.73	16.93	17.67
			2.23	2.25	2.32	2.54	3.37	4.02	4.97	6.06	5.72	5.54	5.15
3.57	3.67	3.73	2.47	2.61	2.68	2.79	3.07	3.90	4.41	4.14	4.18	4.05	4.38
5.04	5.04	4.82	4.99	5.80	6.25	7.67	9.66	13.31	14.70	11.72	8.83	7.34	8.14
2.38	2.09	1.85	1.77	2.16	1.96	2.29	2.76	3.42	3.39	2.62	2.23	2.21	2.49
2.66	2.95	2.97	3.12	3.64	4.29	5.38	6.90	9.89	11.31	9.10	6.60	5.13	5.65
9.19	9.91	10.95	13.24	14.35	15.37	17.63	20.09	22.83	24.55	23.39	22.66	22.77	24.35
4.57	4.97	5.62	7.24	7.96	8.66	9.97	11.53	13.09	13.84	13.00	12.58	12.81	13.76
3.03	3.29	3.55		3.38	3.58	3.96	4.45	4.94	5.38	4.94	4.99	5.44	5.96
1.60	1.66	1.78		4.58	5.08	6.01	7.08	8.15	8.46	8.06	7.59	7.37	7.80
			3.83	3.89	3.87	4.21	4.49	4.98	5.49	5.29	5.09	4.92	5.21
			2.17	2.50	2.84	3.45	4.07	4.76	5.22	5.10	4.99	5.04	5.38

included offices, shops, hotels, places of entertainment, garages, schools and colleges. Speculative work is included, that is, work carried out by contractors on their own initiative for sale. This has contributed to the oversupply of offices throughout the country, particularly in London. In central London alone, empty office space was at record levels at almost 3 million square metres in early 1991 representing an oversupply of 16%. At that time the availability ratio, which expresses space that is, or will become, available over the following 6 months as a percentage of total space, was a record 18%, twice that of 1975. In Docklands, where Canary Wharf is attracting tenants, the ratio was 39%. In 1991, office space in terms of completions reached its peak except in the West End.

Another substantial item of new work is that of housing for the private sector which increased in real terms by nearly 63% between 1985

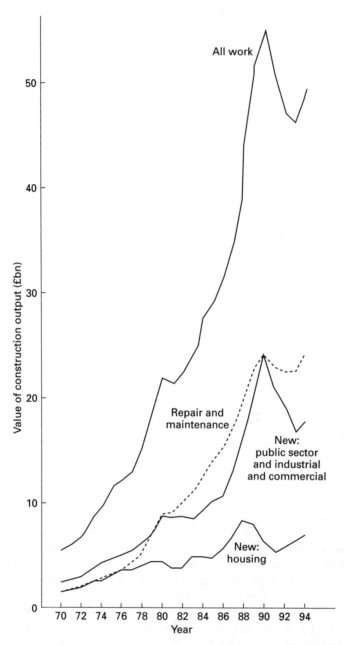

Figure 3.1   *Construction: value of output at current prices in Great Britain*

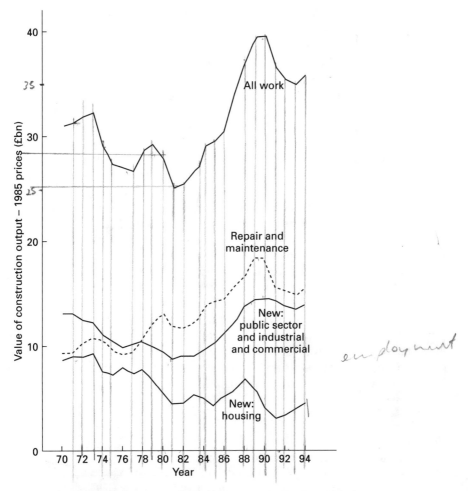

Figure 3.2   *Construction: value of output in 1985 prices in Great Britain*

and 1988 to a peak of £7775m. There has been a fall of some 26% between 1988 and 1994. At the start of the decade housing in the public sector represented 40% of new housing. By 1990 it had fallen to 12%.

Until 1985 the largest individual item of new work was non-housing work for the public sector. A large part of this is made up of expenditure on roads and railways, offices, factories, hospitals and schools. In 1970 it comprised 33% of new construction work. Compared to other sectors, expenditure has fallen in the non-housing public sector during the 1980s from 27% of new construction to 21%.

In real terms, throughout the last two decades expenditure has fallen. This explains, in part, the infrastructure, or crumbling infrastructure, becoming a matter for debate. The new work in the private industrial sector showed decline in real terms until 1987 when orders included the Channel Tunnel project.

A substantial area of expenditure is repair and maintenance work. The £24 544m expenditure in 1990 was largely for housing and highway maintenance. Added weight has been given to the claims of 'the crumbling infrastructure' of the 1980s as repair and maintenance of non-housing public work fell from 32% to 25% of the total repair and maintenance work from 1980 to 1990, although in real terms this is a fall of only 1%. This does not include maintenance work undertaken directly by occupiers on a 'do it yourself' basis. The amount spent on 'do it yourself' goods doubled during the 1980s. It was £3717m in 1990 at 1985 prices.

Table 3.2 and Figure 3.3 show the investment in new construction in terms of the GDP for the countries of Western Europe. Although there has been an improvement more recently when compared with 1980 to 1987, overall Britain invests less in new construction than most other countries in Western Europe. Although these comparisons can be drawn and Britain shown to be acting, apparently, less satisfactorily than much of Europe, the important question is whether an appropriate amount is spent on new construction. Perhaps this can be answered in terms of the needs and expectations of the population. They are that:

Table 3.2   Investment in new construction [3]

| | New construction as % of GDP | | | | | | | | | | |
	80	81	82	83	84	85	86	87	88	89	90
Belgium	14.6	11.8	11.3	10.4	9.6	9.4	9.4	9.3	na	na	na
Denmark	11.8	9.5	9.1	9.1	9.5	9.8	11.1	11.0	10.8	10.4	na
France	14.0	11.7	11.1	10.5	9.9	9.5	10.9	11.1	11.4	na	na
Germany	14.1	13.6	12.6	12.5	12.4	11.2	11.2	10.9	11.1	11.4	na
Great Britain	9.6	8.9	8.9	8.9	9.0	8.5	8.8	9.3	9.9	9.8	na
Ireland	15.0	16.6	14.6	11.8	9.7	8.8	9.1	8.2	7.4	7.6	na
Italy	12.8	11.6	11.3	11.0	11.5	11.0	10.4	9.9	9.9	10.0	na
Netherlands	13.1	11.9	10.9	10.2	10.2	9.4	10.0	10.2	11.2	na	na
Norway	15.3	18.4	16.0	17.8	18.4	15.0	15.4	16.2	16.2	13.3	na
Sweden	12.0	11.4	10.9	10.4	10.3	10.1	9.6	10.0	10.8	11.3	na
Switzerland	16.1	16.7	16.2	16.1	16.0	15.8	15.9	16.4	17.3	18.2	na

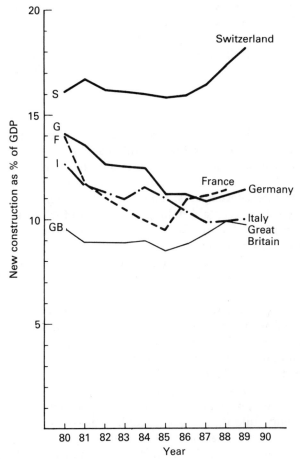

Figure 3.3   *Investment in new construction*

- housing should be of an acceptable standard
- power, water and sewerage systems should not be interrupted
- transportation systems which are reasonably efficient are taken for granted
- health and education are essential and should not be forgotten

That private housing has received greater investment is not in dispute. The government expects that the further privatization of the public services including transportation, will lead to improved efficiency and greater investment during the 1990s. Perhaps, at that stage, investment in the built environment will be comparable to that of other European countries. Nonetheless, at present, Britain's poor economic performance cannot buy the built environment that it ought to have. For example, housing investment alone in Germany is comparable to the overall investment that Britain makes in the built environment.

Product safety

The buildings and other structures in the built environment are subjected to a variety of operating conditions. Their performances are predictable, within limits, and considerable margins of uncertainty are taken into account in the design process. It is not possible to achieve absolute safety against failure of a structure; a level of safety is sought which is acceptable taking into account the sensitivity of the public to structural failure. Occasionally failures occur and were this not the case, there would be a suspicion that construction was depriving other areas of capital expenditure which, if deployed, could provide a far greater return in terms of lives saved. As it is, relatively few structures fail due to inadequate design against loads traditionally catered for and, often, failures are due to ignorance through lack of experience or research. Statistics relating to the frequency of structural failure are few but the probability that a reinforced concrete or steel building will fail by becoming unserviceable or cause unwanted damage or harm is estimated to be annually between 1 in 1 000 000 and 1 in 100 000 000 000 000 [4]. Excluded from this analysis are risks of failure due to design and construction errors; in addition, a degree of care during building use is expected. The risk of actual failure or unserviceability is higher than estimated but, as the development of structural distress is usually protracted, few lives are in danger. This is confirmed by the comparative annual probabilities of death linked with a number of human activities as shown in Table 3.3.

The risk of death from structural failure in the built environment is some ten thousand times less than the risk of death from all causes for a male aged 30; it is much lower than that for all other activities listed in the table. This is at a level that makes activities in the built environment, for example, crossing a bridge, working in an office block, living in a house, stress-free due to the realization that the risk of death from structural failure is extremely small.

New construction

The output of industrial and commercial work of the private sector is about a quarter of the construction work undertaken. Within these

Table 3.3   Annual probability of death [1, 5]

Activity	Hours exposure per annum	Annual risk per 10 000 people
Mountaineering	100	27
Coal mining	1 600	3.3
Car travel	400	2.2
Construction site	2 200	0.8
Air travel	100	1.2
Home accidents	5 500	1.1
Manufacturing	2 000	0.2
Structure failure (built environment)	5 500	0.001
All cases (United Kingdom 1986)		
Male aged 30	8 700	9
Female aged 30	8 700	5
Male aged 50	8 700	55
Female aged 50	8 700	33

figures is contained some of the megaprojects for which the 1980s will be remembered, for example, those at Broadgate, Canary Wharf and the Channel Tunnel. Unless Government policy calls for control of construction in the 1990s, which is unlikely, perhaps future megaprojects of this type will be undertaken, such as the new developments at King's Cross and Spitalfields. It is an area of work in which the property developers have shown both confidence and resources to undertake such massive projects.

Canary Wharf (see Figure 3.4) has been cited as an example of a megaproject [6]. It is the largest single business project in the world. At its peak some 4000 construction workers were employed. When completed it will cover a 29 hectare site and the 26 buildings will contribute more than 1 million square metres of office space and 75 000 square metres of retail and leisure space to London. The main tower block, at 245 metres, is the tallest building in Britain. The estimated 60 000 office workers will be transported by two major road and two major rail schemes which will, for the first time, link the east of London properly to other areas of London and provide a 40 minute access zone. More than £4 billion has been committed to these transport improvements alone.

Figure 3.4 *Canary Wharf was built by the developer for rental*

The Canary Wharf development undertaken by a Toronto-based property owner and developer places emphasis on quality of work together with speed and flexibility of construction. These are important factors in achieving satisfaction from the future tenants and a reasonable return on investment in the construction business. In order to undertake the construction of the buildings more quickly than using conventional techniques, a 'fast track' system (see Chapter 5) was devised. It is an amalgam of techniques which, together, can halve traditional building times. The key to the system is the rigid control of both operatives and materials.

The construction tasks, i.e. steelwork, flooring, concreting, ducting, cladding, etc., follow closely behind each other during construction facilitating the rapid completion of the floors. Strict segregation of the tasks of the services, i.e. lifts, cranes and hoists, is maintained in order to ensure that no work team is without the necessary materials. The logistics of materials delivery is the crucial aspect of the construction work and the problems were compounded by the shape and location of the site. The solution was to transport the bulk of the materials using the river Thames which, unlike the congested road system through the east end of London, is largely free of traffic. An average one barge load is equivalent to five truckloads by road. The delivery of materials was closely controlled with respect to delivery point, date and time. British subcontractors took some time to get used to this close control. Only 70% of the material comes from British suppliers. Materials used came from more than 100 countries and included 10 different types of marble from Italy, stone from Toronto, and cladding panels from Belgium, Canada and the USA. The high quality of materials used in the construction combined with the attractive setting for the buildings in tree-lined avenues, squares and courts which cover 9 of the 29 hectare site, makes Canary Wharf a pleasant working environment,

To carry out the Canary Wharf development when there was already a surplus of office space in London was a high risk enterprise. However, only 10% of the existing office space in London is of a high class in terms of quality of environment and technological efficiency. Canary Wharf has a high quality communications technology designed into the buildings which gives future tenants reliability and flexibility in computer and communications systems; this aspect of commercial business has the highest expenditure after salary and property costs. An added advantage is that technological re-equipping, which determines the future life of modern office blocks, can be more easily carried out than in many existing buildings.

Undertaking large construction projects can lead to significant environmental impact (see next section) both during and after construction and claims of disruption to the lives of local residents. Canary Wharf has not been immune from claims. Local people have launched what could be the biggest group action in English legal history which alleges disruption by dust and noise. It is frequently difficult and costly to prove that the effects of pollution are deleterious to health in such cases and showing unreasonable interference with people's rights to use and enjoy their property may be an alternative approach to effect a successful legal action and compensation. Although most of the benefit associated with a commercial development of this type must go to the tenants, Canary Wharf has considerable benefits for the local residents.

It should not be forgotten that the construction industry has made a substantial contribution to offshore oil/gas exploration and exploitation. An example is the construction and installation of British Petroleum's Magnus field platform and topside modules completed in 1983. The platform is fixed to the seabed in 186 m depth of water at a northerly and exposed location in the North Sea. The platform, which alone stands 212 m high, is fabricated from structural steelwork and is one of a generation of steel jacket structures (see Figure

Figure 3.5   *A new production platform takes shape for the offshore industry*

3.5). The topside structure is essentially a multi-storey refinery which is 75 m square and 32 m high weighing 31 000 tonnes. This is made up of prefabricated modules which are transported to the offshore site and placed into position on the platform. The completed project, valued at approximately £1400m at 1982 prices is fabricated to a high quality specification to withstand the aggressive environmental conditions. In terms of overall size it is of comparable height to and of greater girth than the tower at Canary Wharf.

Environmental impact of the construction project

The impact of the construction industry on the environment is substantial. During the extraction and manufacture of construction materials, their transportation, the process of construction and use of buildings, large quantities of energy are used. Major contributions are made to the overall production of carbon dioxide (see Chapter 4) which exacerbates the 'greenhouse' effect. The environmental impact is global but, during the construction process, communities and individuals are affected as has happened in the vicinity of Canary Wharf.

Society is becoming increasingly concerned with the effect of human activity in the environment [7]. In recent years there has been greater pressure on promoters to state all the likely direct and indirect effects of their projects on the life and amenities of surrounding areas. It had been the practice to supply environmental impact assessments with planning applications for major projects, but these were often not easy to understand as they were expressed in complex, technical language. An EC directive which was adopted in 1985, but did not become law until 1988, now requires parties responsible for both major schemes and smaller projects, if planning officials think fit, to publish environmental assessment statements. This assessment requires a statement of the impact of the project on the surrounding area and details of work proposed which will limit this impact, for example, soundproofing as in the case of a noisy transportation system. To involve the public more closely, the EC directive specifies that the assessment statement be complemented by a statement in non-technical language. A further requirement is that promoters should consider the detailed impact of their project early in the planning process and undertake wide consultations involving the public and environmental groups.

Despite the requirements specified by the directive, there is concern that there is a lack of definition regarding the scope of the assessment and the level of detail required. Consequently, the quality of the assessments varies considerably and an effort is being made by the

Department of the Environment to determine appropriate standards. Assessments tend not to look beyond the confines of the project proposed. Alternative proposals which would compare the environmental factors involved in the choice of a road or rail scheme would be a considerable improvement. The EU is expected to extend the scope of its legislation with a new directive on environmental impact to include the environmental factors accounted for when considering alternatives. A wider consultation process will be required. Exposing unpopular government policy decisions to greater public scrutiny could provoke opposition from national governments when the draft proposals are considered by the Council of Ministers. Should these proposals be adopted, environmental factors might need to be supplemented by political considerations for undertaking projects in particular industries or geographical locations where employment and use of resources are thought desirable.

Refurbishment

The annual programme of new housebuilding contributed some 200 000 new homes to the building stock in Britain in 1989. In contrast, 6684 were demolished or closed through programmes of slum clearance. Yet refurbishments were undertaken on 414 049 homes and represented a considerable undertaking in terms of construction capacity. Refurbishment was also carried out on many larger buildings including office blocks, departmental stores and industrial buildings. As the yearly rate of new construction is only a little more than 1% of the existing building stock, it is estimated that the industry is able to effectively refurbish and adapt existing buildings.

Most buildings are constructed from materials which exceed the practical life of the structure and refurbishment makes good use of this inherent extra value. A professional refurbishment gives the building owner a redesigned, modernized building which fulfils all requirements without many of the disadvantages of a new building. The result is a measure of conservation although this is not the primary objective. Rather, refurbishment represents a recycling concept which prevents potentially valuable building stock from being prematurely wasted [8]. This is not to say that refurbishment is necessarily the answer to a building owner's requirements. Industrial buildings and office blocks do not always lend themselves to refurbishment. The pace of change has accelerated and building needs change rapidly. Dated building shells may be unsuitable due to physical limitations to house intelligent buildings of the future even with expensive adaptations and upgrading to give greater flexibility and

meet information technology specifications. Although extensive reno-
vations and refurbishment of large buildings may cost much more
than that needed to provide a completely new building, this is not
likely to be so in the case of houses. For these structures refurbishment presents an attractive solution. However, this is the sector of
construction most likely to receive the attentions of the cowboy
builder (see p. 68).

Repair and maintenance

Repair and maintenance has become a substantial area of work. At
the beginning and end of the 1980s, repair and maintenance made up
some 41% of all work. During the mid-1980s it peaked at 46%. This
contrasts with the early 1960s and early 1970s when repair and maintenance was some 20% and 30%, respectively, of all work. This is not
due entirely to the volume of new work being smaller. In real terms,
repair and maintenance exhibits an in-phase cyclical variation with all
work. Nonetheless, it has grown by some 54% since 1972. Elsewhere
in Europe, repair and maintenance varies from 14–44% of all work [9].
Reports from the USA suggest that this activity forms a much smaller
proportion of construction output than in Britain [10]. Perhaps Britain
is a 'make do and mend' society whereas in the USA there is a greater
tendency to 'demolish and start again'. Decisions may be influenced
by material prices. For example, the relatively low price of steel in the
USA would suggest that this is a favoured alternative when choosing
between concrete and steel as a structural material. Consequently,
there is scope for the better maintenance of quality. There is also a
greater tendency to fabricate construction components under factory
conditions in the USA and erect them on site. Under these circumstances, quality should be easier to maintain, although this is not
always the case.

Repair and maintenance is also linked to R&D. Due to insufficient
R&D there have been problems with the durability of concrete; precast multi-storey flats, and foundation problems due to shrinkable
clays during periods of drought.

Quality of the construction product

Despite the fact that the risk of death or injury due to failures in the
built environment is small, the quality of the building product often
leaves a lot to be desired. Much building is for buying and selling.
When the property under construction is to be owned by the developer, standards are generally better. It is an industry in which:

- there has never been a requirement for the work force to be formally qualified and skills are generally developed through time serving
- much of the work is carried out by subcontractors in a climate in which some 50 firms come into existence every day and a similar number go into liquidation or become bankrupt every day
- there is a paucity of R&D involving new materials, designs and techniques
- there is often poor management and supervision

It is not surprising that construction products take too long to build, require remedial work which is sometimes extensive, deteriorate rapidly, do not meet the user's requirements and cost more than they should. Surprisingly, the owner of a new house finds that there are no common law rights as to the sufficiency of his new property. Instead, in the light of the faults found in the product (studies have shown that about 50% of faults originate in the design office, about 30% on site and about 20% in materials) which the owner would like put right, there is an uphill battle involving the National House-Building Council, the builder, if he is still in business, and the courts. It is often difficult to obtain redress due to poor materials, design and bad workmanship. Justice through the courts can take years with the worry that the case may not be proved sufficiently beyond doubt as to who is the offending party. The fragmentation of the construction process brought about by the several separate contributors to it encourages recourse to the law.

Although not specifically intended for the purpose, a measure of quality has been imposed on the construction professions which are responsible for structures through the Latent Damage Act of 1986. Action can be taken in the case of proven negligence up to 15 years after the event by those affected. In 1990, over 3500 firms were being sued for claims [11] which is in itself an indication of the level of quality achieved by the industry.

Codes of Practice and other standards are important instruments developed over many years which are designed to provide guidance for the industry. These codes are subject to review and revision as a greater knowledge of techniques and materials becomes known. For example, a 1992 report [12] suggested that standards of insulation for housing in Britain were grossly inadequate and Building Regulations must be revised. Nonetheless, some faults are due to insufficient R&D. Examples of distress have involved the design of foundations on soils subjected to shrinkage. Buildings over much of the southern part of England have had settlement problems, some requiring extensive underpinning; perhaps more underpinning has been carried out than has been needed. Annual subsidence claims have increased from £5m in 1975 to £540m in 1991. Other examples include the deterioration of concrete and subsequent failures. Industrialized building has not been successful in Britain due to a combination of poor design and insufficient supervision of the components' fabrication leading to erection difficulties. Motorways in Britain require extensive remedial work although this is said to be due, in part, to the operational requirements which are now more demanding than the specifications for the original designs.

A survey undertaken by the National Economic Development Office (NEDO) in 1978 investigated the opinions of the construction industry held by clients who had had recent experience of the work of the industry [13]. Of the 241 clients, 217 were reasonably satisfied with the design and planning stage, 214 with the construction stage, 168 with defect rectification, 203 with final cost and 184 with total time taken. The other firms in each case were either slightly or very dissatisfied. Clients complained of poor management with contractors having insufficient control over subcontractors. Some felt that the construction industry never really did a satisfactory job because it was badly organized.

An analysis of error [14] isolates a common pattern for defects and classifies them by:

- ingress of rain
- condensation
- cracking, breaking, movement and detachment of components and even the structure itself

An analysis of reported shortcomings [15, 16] shows that water ingress is reported in 177 out of 599 case studies, i.e. 30%, and component shortcomings in 297 out of 599, i.e. 50%. Reference [16] suggests that the biggest contribution to design faults is by choice of the wrong materials or components; a contributory factor is the failure to understand movement and vapour controls. An investigation [17] into prefabricated Airey Houses, of which 26 000 were erected during the post-war years, showed that design oversights allowed ingress of water to the base of the columns. The resulting microclimate was hostile to the reinforcing steel which caused degradation of the columns. It was judged that the design faults were brought about by an ignorance of the fundamentals of building science; fundamentals often not understood in the design of present day structures.

Few owners are fully satisfied with their new houses. In 1982 the results of a three year survey carried out on 12 public sector and three private sector housing sites [18] showed that 955 different kinds of faults were found. In total, there were some 72 000 faults in the 1725 dwellings inspected; on average, about 42 faults per dwelling. All these faults were capable of giving rise to inferior performance. About one-third of the fault types found were unlikely to be remedied because of the high cost or impracticability of the remedial work necessary. Overall, it was judged that the extent of poor quality is considerable and major improvements to the control of quality are needed. Without such improvement, modern homes will require excessive maintenance and repair. This could give rise to costs which will be comparable to two-thirds of the initial cost.

Table 3.4 indicates the building element in which the faults occurred. Two-thirds of these faults are in the areas of strength and stability, durability and weathertightness.

In 1984 the Department of the Environment investigated repairs and improvements which had been carried out to housing. The cowboy builder was blamed for the escalating costs of refurbishment and the poor durability of the work undertaken. The construction industry has a poor reputation for quality control due largely to a number of interrelated reasons. Supervision and management is often poor; good practice is not well communicated throughout the industry as much of it is distant from up-to-date technical information; there is a reluctance to investigate error and feedback experience; it is difficult to motivate the workforce to undertake the work to a good standard. Finished work can be quickly covered up and faults are found only if

Table 3.4 Building elements in which the faults occurred [18]

Element of building	% of total kinds of faults
External walls	20
Roofs	19
Windows and doors	13
Floors	11
Services	9
Substructure	7
Internal partitions	4
Separating walls	4
Stairs	4
Planning and layout	4
General and external works	4

the work proves to have considerable shortcomings during use. As there is little political will to introduce change and the rights of the consumer are few, it is likely that the construction industry will continue to produce a poor quality product.

Influence of resources on construction activities

There has been a number of studies undertaken [19, 20] to investigate the effect of resources on the types and stages of construction work, and in addition, to find the extent that the construction industry as a whole can respond flexibly to changes in demand.

The findings of the investigations can be summed up as follows:

- The numbers of professionals working in the industry who undertake the design process tend not to change rapidly due to upturns or downturns in demands. Substantial changes can be made to their rate of working. In contrast, the numbers of technicians change more rapidly.
- There is a serious waste of resources during a downturn in demand and a risk of a decline in quality during a rapid upturn.
- The maximum possible rate of continuous real growth of turnover for an average individual contractor or subcontractor in which growth is expected is estimated at about 10% per annum. A large and sustained expansion in growth for some years would only be possible if most contractors and suppliers have confidence enough to plan, train and invest for the anticipated sustained growth.
- In terms of numbers there can be rapid change in the number of craftsmen available. However, this is achieved because of the acceptance of untrained and inexperienced men in times of growth and loss of skilled men and apprentices when a downturn takes place. A rapid expansion of work leads to wage inflation and a loss of quality; contraction causes unemployment and wasted training.
- The flexibility in materials supply was believed to be some 10% increase per year in the event of an upturn in the industry. It is important not to consider the construction industry home suppliers in isolation. The rapid upturn of the late 1980s showed that the increase in material supply could considerably exceed 10%; this was at the expense of a significant international balance of payment deficit on construction materials (see pp. 85–86).
- Shortages of plant are rarely cited as a major problem for the building and civil engineering industries and are unlikely to be a major constraint on construction work. This may be due to the way in which construction work is undertaken; a later (p. 251) section

suggests that less reliance on plant and equipment is partly responsible for low productivity in the British construction industry.

Future programmes

The factors influencing future demand for construction are many. They can be interdependent and conflicting. They include:

- political decisions
- economic and financial factors: the general level of economic prosperity or otherwise including changes in the GDP, interest rate and investment climate
- social requirements and expectations: public or political demand for improved services or infrastructure
- changes in, or movement of, population
- technological developments: exploitation of new resources, materials and methods

It is expected that the 1990s will be the decade in which much of the infrastructure, for example, transport and water facilities, will be repaired and replaced. Although it is projected that the total output of the construction industry will fall by some 4% in 1991 and 3.5% in 1992, it is expected that the newly privatized utilities and the road programme will receive more substantial investment. Examination of past records suggests that construction is the lead sector of the economy and, when recession has forced down construction prices far enough, a recovery in the demand for construction leads to an upward demand in trade generally. The recovery in construction is led by an increase in house building and the end of recession can be determined by examining house completions, which increase spectacularly. A surge in house building is unlikely to take place for some time. Only civil engineering will escape the worst of the present recession.

The construction product for the 21st century should be geared to providing good quality homes and associated services, an efficient transportation system, health and education infrastructure, commercial and industrial premises. Government expects private investment to provide the bulk of the investment. During the next decade, the population of Britain will grow by nearly 2m. The bulk of this increase will be in the over 50 age group which will place greater demands on a proportionally smaller working population.

Housing

In 1988, 201 928 new homes were completed; 173 361 for the private sector and 28 567 for the public sector and housing associations, etc. After 11 200 conversions and 8415 losses due to slum clearance and other factors were taken into account, the net gain was 227 891 homes. Some 414 000 dwellings were renovated. Based on the number of new homes completed, the average unit costs for private and public sector homes were £45 000 and £35 000, respectively. It is surprising that the difference is not greater. In terms of new building there has been a substantial decline in home building since the peak in 1968, when 222 000 private sector and 192 000 public sector homes were completed. The proportion of housing occupied by owners has risen steadily throughout the 1980s to 65%, while 25% is rented from local authorities and 10% privately rented.

On average, the population of Britain appears to be relatively well-housed with some 22.8 million homes available for a population of 55.5 million; 2.4 inhabitants per home. Nonetheless, there are shortcomings in the state of the housing stock. Every 5 years the Department of the Environment (DoE) commissions a report which assesses the state of private and public housing. The 1986–1987 survey estimated that:

- 900 000 houses were unfit for human habitation (this includes structural instability, serious disrepair, dampness prejudicial to health)
- 460 000 homes lacked basic amenities
- 2.4 million homes required urgent and essential repairs costing more than £1000 each

Slum clearance in Britain declined significantly during the 1980s but has recently increased by 6%. The census indicated that there was a surplus of over 500 000 dwellings for the whole of Great Britain. However, this does not take into account that the distribution of available housing throughout Britain does not correspond with the distribution of families requiring household spaces. It is estimated that 60 000 to 90 000 homes must be built each year to meet needs at the low cost end of the market, but by the end of the 1990s a shortfall of a third of this estimate is expected.

Housing investment during the 1990s has been the lowest since the end of the Second World War: 35% of the housing stock is pre-1940, 34% was constructed during the immediate post-war years with the remaining 31% being constructed during the past thirty years. In 1996, it was estimated that there were over 800 000 empty houses that would require in excess of £70 bn spending on them to bring them up to acceptable standards. At the same time there are estimated to be over 360 000 homeless persons, changing family patterns and life longevity that require considerable investment in new houses over the next thirty years.

Health services

The new construction work for the health services is shown in Table 3.5 and Figure 3.6.

Table 3.5   Contractor's output for health services; new work excluding infrastructure [1]

					£m						
Year	84	85	86	87	88	89	90	91	92	93	94
Public sector	500	571	569	542	550	731	853	836	956	869	817
Private sector	99	114	125	153	165	272	374	327	344	300	281
Beds	411	405	394	373	358	343	330	313	295	287	280

The actual spending has increased, more significantly in recent years, although there has been a recent decline. A further statistic, the average daily number of available beds for in-patients, is also worthy of consideration. This statistic may be unrelated to the building programme and depend on staffing levels. Nonetheless, there is decline. No doubt, private investment is believed by government to be the answer to meeting the needs of an ageing population.

Education

Table 3.6 and Figure 3.7 show the public sector investment in new construction for schools and universities in actual and real terms.

Public sector investment has recently increased considerably in both schools and universities. In the latter case it is likely to continue to rise to cater for the participation rate of one in three of all 18–19 year olds entering higher education by the year 2000 [21]. This is equivalent to nearly doubling the number of full-time-equivalent students between 1987 and the year 2000. Table 3.6 and Figure 3.7 indicate that the necessary requirements, in terms of investment in new building stock, are being satisfied to set about achieving this goal.

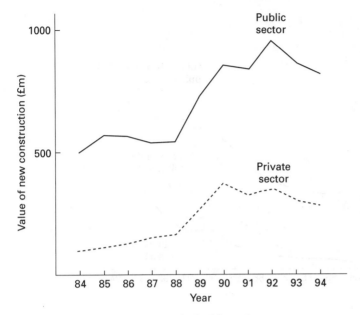

Figure 3.6   *Contractors' output for health services*

Transportation

Goods transported in Great Britain have increased from 1877 to 2145 million tonnes during the past 10 years; an increase of 14.3%. This increase has been taken solely by the roads and amounts to some 31 thousand million tonne kilometres; equivalent to 41 000 extra goods vehicles. The total increase in all motor vehicles during the corresponding decade has been 5 544 000 or 31%. Some 23 302 000 motor

Table 3.6   Contractor's output for schools, colleges and universities; new work excluding infrastructure [1]

						£m					
Year	84	85	86	87	88	89	90	91	92	93	94
Public sector:											
Schools and											
colleges	327	302	320	344	389	486	577	576	729	735	780
Universities	29	38	43	56	62	94	164	151	226	345	407
Private sector	64	71	80	83	145	209	204	206	195	170	120

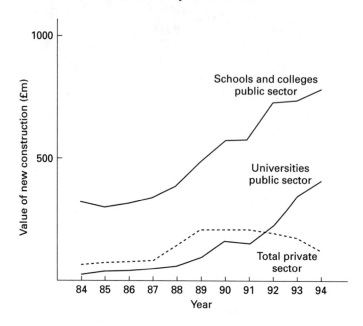

Figure 3.7 *New construction for schools and universities*

vehicles now use the roads of Great Britain. In order to cater for this traffic there are 354 315 kilometres of public roads in Britain, 50 411 kilometres of which are motorway, trunk or principal roads. The total increase in public roads during the 1980s has been 5% but only 2% increase has occurred in the higher-grade roads. During the past few years the investment in new roads has risen from £894m in 1984 to £2154m in 1994 (see Table 3.7 and Figure 3.8). These figures do not include investment for repair and maintenance.

It is scarcely surprising that the roads of Great Britain are becoming more congested. Only 20% of goods are transported by other than road transport and the contribution of alternative transportation has remained static at some 387 million tonnes for a decade. In the case of rail, the goods transported have dropped by 21 million tonnes. An

Table 3.7 Contractor's output for new road construction [1]

	Public and Private sectors										
Year	84	85	86	87	88	89	90	91	92	93	94
£m	894	929	964	972	1271	1448	1695	2102	2023	1951	2154

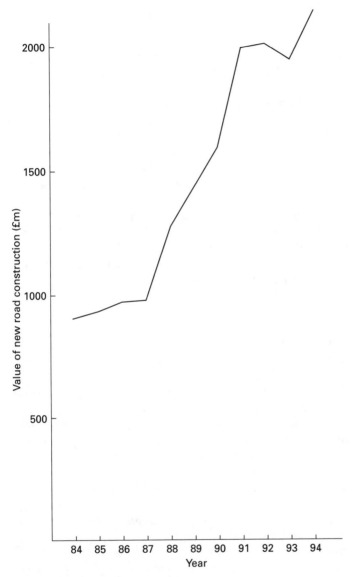

Figure 3.8   *New road construction*

investment programme with a policy which closely considers transportation alternatives to prevent greater overcrowding of the roads
would seem to be essential.

References

[1] Central Statistical Office. *Annual Abstract of Statistics.* HMSO

[2] Department of the Environment. *Housing and Construction Statistics.* HMSO, London

[3] *UN Annual Bulletin of Housing and Building Statistics for Europe*

[4] Rationalization of safety and serviceability factors in structural codes. *CIRIA Report 63.* 1977

[5] Bulson, P.S., Caldwell, J.B. and Severn, R.T. *Engineering Structures.* University of Bristol Press, 1983

[6] 'The biggest building site in Europe' in *The Daily Telegraph Magazine.* 1990

[7] 'Minimum impact. Environment' in *New Civil Engineer.* 1991

[8] Marsh, P. *The Refurbishment of Commercial and Industrial Buildings.* Construction Press, London and New York, 1983

[9] Murray, S. 'Structure of the European Construction Industry – an overview', *European Construction Industry Analysis Conference.* London, 1989

[10] *UK and US Construction Industries. A comparison of design and control procedures.* RICS, 1979

[11] Finniston, M. 'Engineering in the 21st century'. Maitland Lecture. Institution of Structural Engineers. London, November 1990

[12] RIBA Environment and Energy Committee. *Supply and demand in a greenhouse economy.* May 1992

[13] NEDO. *Construction for industrial recovery.* HMSO, London, 1978

[14] Kennaway, A. 'Errors and failures in building – why they happen and what can be done to reduce them' in *International Construction Law Review,* **2,** 1984

[15] Bickerdike, A. 'Design failures in buildings' in *The Builder.* 1971–76

[16] Freeman, I.L. *Building failure patterns and their implications.* Building Research Establishment, 1975

[17] McArthur, H. *Defects in Airey Houses. Structural Survey.* 1984

[18] Building Research Establishment. *Quality in traditional housing.* HMSO, 1982

[19] NEDO. *Construction into the early 1980s.* London, 1978

[20] NEDO. *How flexible is construction. A study of resources and participants in the construction process.* London, 1978

[21] *Higher Education. A new framework.* HMSO, 1991

4

Materials and components

How can modern materials be part of a building until all those involved in design, construction and maintenance understand them?

Frank Lloyd Wright, 1932

Construction materials

The construction industry uses materials from many different sources. These include materials produced by continuous mass production techniques under factory conditions, for example, cement, structural components, bricks; materials and components produced on site, for example, concrete; and natural products such as soils and timbers. It serves to illustrate the complexity of the construction process that a multiplicity of components and services are drawn from different industries, some of which do not consider themselves to be part of the construction industry, before the construction project, often a one-off job, is undertaken and completed. The construction process is not clearly defined as many building-materials producers may not be included. For example, construction firms may use up to 20 000 different items from materials stocks yet the industry may be a less than 20% user for hardboards, strip sheets and polyester resins.

A wide range of different materials is used for constructional purposes and the annual expenditure on materials was some £20 billion by the British construction industry in 1991. There tends to be a lack of reliable data on the amount of material used, but it accounts for about 40% of all construction costs. In the case of repair and maintenance the use of materials may be as little as 15% of the total output cost. For work involving a large proportion of expensive components

Table 4.1 Production of building materials and components: (a) United Kingdom; (b) including sand used in the production of sand lime bricks; (c) from 1979, volume sold; (d) volume sold [3]

	Unit	1970	1971	1972	1973	1974	1975	1976	1977	1978	1979
Building bricks (excluding refractory and glazed)	millions	6 062	6 541	6 938	7 183	5 575	5 046	5 406	5 067	4 842	4 887
Cement(a)	thousand tonnes	17 171	17 697	18 048	19 986	17 781	16 891	15 780	15 457	15 916	16 140
Building sand(b)(c)		18 450	19 743	21 314	23 619	20 295	21 444	20 431	18 608	18 510	18 983
Concreting sand(c)		33 523	33 329	33 147	37 587	32 468	32 667	31 118	27 483	29 164	29 455
Gravel and concrete aggregates(c):		57 331	58 754	62 873	67 943	59 876	63 069	58 405	53 138	54 426	54 056
Crushed rock aggregates(c):											
Roadstone		na	na	46 132	74 554	67 557	58 352	52 125	50 130	51 717	56 198
Concrete aggregate		na	na	18 705	21 403	16 958	16 930	16 227	15 259	15 872	15 588
Fibre cement products		na	na	na	na	na	na	na	450	429	406
Concrete pipes (a)(d)		988	1 077	1 271	1 470	1 330	1 424	1 374	1 111	1 158	1 063
Plaster		868	942	1 019	1 125	1 038	1 047	1 053	988	975	949
Plasterboard	thousand m²	82 118	95 401	103 241	112 801	110 861	104 396	116 999	101 827	114 120	113 961
Concrete building blocks(a)		49 817	57 767	68 414	72 777	54 542	62 137	69 763	66 866	71 060	73 090
Clay roofing tiles(a)(d)		953	969	1 151	1 238	1 195	1 108	1 170	1 192	1 207	2 265
Concrete roofing tiles		22 349	27 787	30 144	30 669	27 359	25 942	29 839	24 151	27 899	28 263
Ready mixed concrete(a)	million m³	23.3	25.2	27.2	31.7	27.8	26.7	24.5	23.5	23.8	24.4
Rods and bars for reinforcement	thousand tonnes	na	na	na	na	na	na	na	948	1037	982

Table 4.1 cont.

1980	1981	1982	1983	1984	1985	1986	1987	1988	1989	1990	1991	1992	1993	1994
4 562	3 725	3 517	3 806	4 012	4 100	3 971	4 222	4 682	4 654	3 802	3 212	3 000	2 639	3 114
14 805	12 729	12 962	13 396	13 481	13 339	13 413	14 311	16 506	16 849	14 740	12 297	11 006	11 038	na
18 005	15 675	17 044	19 151	18 187	18 346	20 423	20 620	23 415	23 209	20 948	18 079	16 769	17 406	17 168
26 699	25 427	25 242	26 720	28 172	28 694	30 347	32 823	39 174	41 223	37 213	31 239	28 573	28 021	27 986
51 454	48 351	48 920	54 929	53 316	54 564	54 729	58 185	67 566	66 718	58 010	48 598	43 557	44 043	43 521
57 262	49 128	52 998	57 084	54 084	59 356	63 424	73 383	83 047	89 748	88 172	87 135	80 118	81 650	na
13 653	11 205	12 721	14 582	14 360	13 214	13 715	15 443	17 978	19 326	1 880	15 203	14 729	15 786	na
352	266	262	240	256	253	217	205	251	221	235	134	121	129	134
692	648	694	744	701	571	630	533	700	717	737	na	na	na	na
960	775	764	855	912	876	912	na	na	na	na	na	na	na	na
111 132	101 987	107 867	120 819	127 668	122 599	134 501	na	na	na	na	na	na	na	na
66 820	55 565	63 685	75 642	81 643	74 445	87 154	96 997	110 029	107 999	91 154	74 631	68 194	74 287	87 548
1 698	1 635	1 632	2 206	2 051	2 143	2 587	3 019	3 459	3756	2 965	na	na	na	na
28 813	23 345	25 551	33 243	34 685	27 870	30 843	34 707	38 818	35 787	31 510	26 359	21 490	24 574	28 149
22.4	19.9	20.7	21.5	20.8	21.6	21.5	24.4	28.8	29.6	26.8	22.5	20.8	20.8	22.9
709	711	675	814	846	916	879	914	1168	1219	1241	na	na	na	na

the proportion of materials can be as high as 60% of the cost [1]. Perhaps it is not surprising that the building materials industry has been shown to operate anti-competitive trading practices. It has been investigated by the Office of Fair Trading and a 10-year inquiry into the ready-mixed concrete industry is continuing. Seven of the largest suppliers have admitted taking part in secret price fixing and market share cartels [2]. Other cartels uncovered have involved steel reinforcing bar and float glass suppliers.

Table 4.1 and Figures 4.1 and 4.2 summarize the output figures for some of the more important building materials and components produced in Britain during the past two decades [3]. These are based on returns made by manufacturers and producers. They represent total production and are not merely the quantities available for construc-

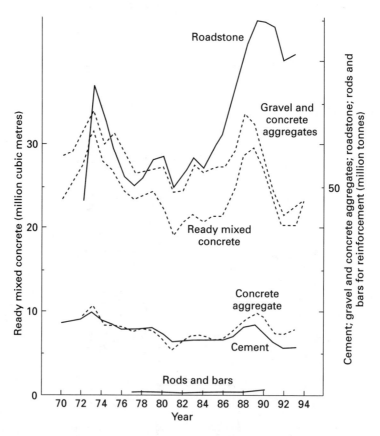

Figure 4.1   *Production of building materials and components*

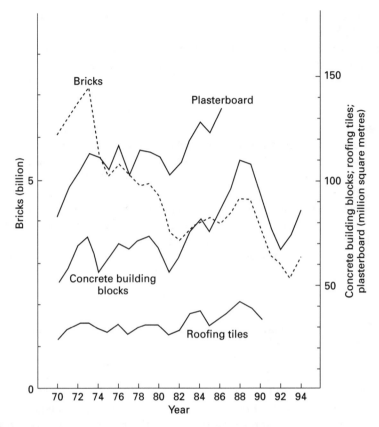

Figure 4.2 *Production of building materials and components*

tion purposes. A study of the utilization of materials in construction [1] indicates that the traditional constructional materials, for example, aggregates, bricks, plaster, steel, glass, timber and joinery account for rather more than 50% of all materials, by value, used in construction. Materials and components for services account for a further 25%. See Table 4.2.

Until fairly recently construction materials have been selected largely on the basis of their structural integrity with particular reference to strength, stiffness and durability. Greater attention is now paid to health and environmental issues; the latter issue includes the energy requirements of the industry.

A considerable amount of energy is required to provide and transport the materials used. For example, the materials needed to construct a typical three-bedroom detached house will use, in their production, an amount of energy equivalent to that consumed by heat, light and power during 2–5 years occupation of the house. The process of the materials' production will generate over 20 tonnes of carbon dioxide [4]. The most energy-intensive materials, for example, paints, metal and glass, tend to be the most expensive materials; these are used most sparingly in house construction. Cement, plaster products and bricks have lower unit costs but, because of large volumes used, form a greater part of the cost of a house.

The production of energy in order to convert the raw materials into acceptable construction materials is environmentally damaging. Some European countries, notably Germany, use a labelling system for construction products which indicates which of them are low energy products and, hence, environmentally friendly. The German system is known as the 'Blue Angel'. Traditional materials such as wood, brick and concrete offer structural efficiency at low energy costs. The following examples indicate the relative amounts of energy required to produce structurally comparable materials [5]:

- A light-gauge cold-rolled steel purlin compared with a 300 mm × 50 mm rough sawn joist of the same stiffness has nineteen times more energy cost. Even if the timber is planed and treated, the steel member requires five times the energy cost.
- A 305 mm × 165 mm steel I-beam has the equivalent performance of a 550 mm × 135 mm glulam softwood beam, but needs six times the energy. The comparable 400 mm × 250 mm reinforced concrete beam has five times the energy cost of the glulam beam.
- Comparing the overall cost in ground floor construction for a typical domestic house based on a particle board/timber joist/timber bearer/concrete wall and footing solution with that of a solid mesh reinforced concrete floor/moisture barrier/support walls/footings showed that the energy consumption of the latter was almost double that of the former.

These examples indicate that there is an energy factor to be considered in the choice of construction materials. However, the full considerations of energy factors are complex and far-reaching as materials costs cannot be considered in isolation. They involve not only the choice of materials but further considerations including construction, lifespan and demolition. At the present time the construction industry often chooses durable materials based on first cost and availability. It is unlikely that significant change will take place barring legislation or cost advantages.

It should also be remembered that, although the materials chosen are ostensibly durable, the implementation of poor design and con-

Table 4.2   Estimated relative significance by value of construction materials [1]

Structure 40%	Aggregates, at site – 8%
	Cement, sent to site – 3%
	Ready mixed concrete – 7%
	Manufactured concrete products – 3%
	Bricks and blocks – 8%
	Steel – 7%
	Metal windows and doors – 2%
	Glass – 1%
	Insulation – 1%
Timber Trades 20%	Carpentry – 13%
	Joinery – 5%
	Ironmongery – 2%
Services 25%	Heating, ventilating and air-conditioning – 12%
	Plumbing – 7%
	Electrical – 5%
	Lifts – 1%
Finishes 5%	Plaster and plasterboard – 1%
	Paint – 4%
Other 10%	Iron not elsewhere specified – 2%
	Plastics not elsewhere specified – 3%
	Asphalt, bitumen and roofing felt – 5%

struction techniques causes their premature degradation and leads to defects for which remedial work can be expensive. For example, it is estimated that corrosion costs some 4% of the GDP of Britain. Good practice is a basic requirement of design and construction in order to ensure that materials are adequately protected to maintain their effectiveness throughout the lifetime of a structure.

Chapter 3 indicated the main divisions between different types of construction work. New work has accounted for some 60% of construction during the past few years. In new work, structural materials are primarily required in the first 60% of the contract period whilst materials and components for services and finishes are used more predominantly after the completion of 40% of the contract. Thus, variations in the amount of construction work undertaken due to economic pressures influences some manufacturers and producers more immediately. Whether or not the manufacturers can respond appropriately reflects on the flexibility of the construction industry and is worthy of examination.

Capacity and flexibility of supply

Construction work can be divided into four main market sectors. During 1987/1988 construction work expanded by over 17% and the market shares of each area based on value of output are shown in Table 4.3.

Table 4.3   Sensitivity of materials requirements to changes in demand for construction work

	1987 %	Contribution to expansion %
New housing	20	5.0
New: public sector	11	1.3
New: private sector	24	6.3
Repair & maintenance	45	4.6

In order to achieve over 17% expansion, there was an expansion in all the main market sectors. The contribution of each sector to the overall expansion is shown in Table 4.3. As housing accounts for 61–75% of the bricks used in construction [1], a 5% contribution to expansion, which is a 25% rise in housing, demands some 16–17% increase in bricks supposing that housing does not change its mode of operation. When the contributions in demand for bricks from the other market sectors are added to the 16–17% it is clear that the overall increased demand is substantial. Similar conclusions can be drawn for other materials.

Construction is the only industry using bricks whilst some materials used by construction are used also by other industries. For example, construction uses about 50% of the heavy steel sections required by industry. If there is an increased demand by construction the producers are partially cushioned from large changes unless the non-construction industries are expanding at a similar rate to construction. Generally, there is a measure of flexibility unless construction is the sole user of the material.

A number of factors influence the response that manufacturing and production organizations can make to meet changing requirements in demand. The primary factors are stock levels, changes in the working rate of existing capacity, alterations to capacity and available transportation. Secondary factors include product substitution, changes in imports/exports and changes in the total market's use of materials. Stock analysis indicates that the maximum difference between the peak and trough of most materials stock is two weeks to two months production. For example, the Brick Development Association has recommended to its members that they should stock two months supply. Further stocks are also held by builders' merchants and contractors. There are limitations to stock carrying which are imposed by space, cost and deterioration of the product. Sustained upturns or downturns in demand can be met by changes in the working rate of existing capacity, investment in new capacity or the closing or mothballing of existing capacity. However, closures and mothballing can reduce competitive efficiency if delivery distances increase excessively and is avoided wherever possible. Spare capacity is normally 10–20% for most materials. New investment is more difficult to undertake in cases where a high capital investment is necessary. Where the continuous operation of a kiln is required, re-opening or closure is a major operation. A considerable investment in new capacity for cement or bricks is required in order to develop a modest increase in production and producers need confidence in the level of future demand before investing in new capacity. Planning permission is an important factor in the decision to increase capacity, particularly for cement and aggregrates.

Table 4.1 shows that brick production between 1970 and 1994 varied between 2 639 million and 7 183 million bricks; a variation in production of some 63% of which 25% took place between 1982 and 1988. The 25% variation corresponded to the significant increase in housebuilding which is the dominant factor in the demand for bricks. Cement production shows relatively lower variation in production; since 1970, some 44%, with 21% taking place between 1982 and 1988.

For some materials, the response to variations in demand depends, in part, on the availability of transportation. This is the case for materials which are of low unit cost, yet bulky and heavy, for example aggregates. Although there was a fear that a rapid upturn in demand of more than 10% in one year would lead to haulage difficulties, this has not proved to be the case. Table 4.1 shows that increases in volume of bulk materials sold have, in some cases, significantly exceeded 10% per year.

Bricks, cement, steel and many other building materials, particularly timber, need to be imported in order to meet the demands of the construction industry. Some other materials are exported. The exports and imports of materials and components used in construction were equivalent to an estimated 17% and 27%, respectively, of total construction material usage in Britain in 1976 [1]. This corresponded to a deficit of £471m. During the construction boom of the late 1980s the balance of trade deficit in building materials was nearly £3000m or

£1150m at 1976 prices. This represented 12% of the total deficit in manufactured and non-manufactured goods [6]. It demonstrates that the expansion of the construction industry cannot be carried out in isolation to the rest of industry. It further serves as a reminder that the performance of the construction industry as a whole is not inspiring.

Finishes

Examination of Table 4.1 shows that during the past two decades the use of plasterboard has increased by over 30% whilst that of plaster has decreased by almost 10%. This reflects the trend, which is more emphatic outside Britain, to reduce the wet trades to a minimum, leading to faster construction and reduced labour input on site. In addition to reducing plastering, floor screeds are also eliminated where possible. For example, hardwood flooring squares are laid on a thin polystyrene sub-base to take up undulations in the power float finish [7].

Hazardous materials

It is essential that construction work is carried out with materials that are environmentally as well as structurally safe. Yet, in 1985, it was estimated [8] that much paintwork in housing is still heavily leaded, everyday drinking water is received in many homes from old leaded pipes, 4 million of Britain's 4.5 million council homes are likely to contain asbestos as are 80% of metropolitan schools/colleges and 77% of social service buildings.

The following materials are considered to be hazardous [8]:

- Asbestos and asbestos-cement products: these have been used in buildings for roofing slates; troughed sheeting; boards; rainwater pipes and gutters; glazing compounds; insulation board and laggings. Asbestos is released during weathering, wear, maintenance and disposal which causes it to be a significant health hazard. This is also the case for asbestos fibre/bitumen products which have been used for roof flashings or in bitumen roofing felts. Vinyl and thermoplastic floor tiles may contain asbestos fibre which can be released with wear and ageing.
- Lead: this has been used for roof and pipe flashings; leaded lights; paints; pipework. The latter application is a major hazard in soft water areas and is not now specified for use. It has been estimated that 45% of households in Britain use water that has passed through lead pipes or tanks although it is not often in concentra-

tions above 10 microgrammes per litre. A random sample taken in 1975/76 showed that about 10% of households in England and Wales, and over 55% of households in Scotland, have quantities of lead in drinking water that exceeded 50 microgrammes per litre which is the maximum permitted upper limit set by the EU Directives. Lead solders used for pipe jointings are also hazardous. Lead-based paints offer an improved durability when compared with other paints but, if they contain greater than 5000 ppm, the lead released by abrasion and decay is a hazard when ingested by young children. This is also the case for varnishes and wood stains with more than 5000 ppm total lead.

- Vermiculite: this has been used for roof and boiler insulation which, if it includes asbestos fibre dust, is hazardous. Non-fibrous vermiculite can safely be specified for insulation.
- Polystyrene and polyurethane foam: these have been used for insulation. Despite a slight risk from the latter, they are generally satisfactory if undisturbed but, during fire, there is a health hazard resulting from toxic fumes.
- Adhesives: these are used for bonding materials and when cured present no significant risk. However, when fresh, adhesives are an irritant and working spaces in which adhesives are used need appropriate ventilation.

Although many of the above components which contain hazardous materials are not now used in construction work, there are, nonetheless, existing buildings that contain these materials.

Decisions regarding the removal or otherwise of materials are not always straightforward. In the case of materials containing asbestos fibres, a good deal of preparatory work, for example sampling, is necessary before removal and disposal which can be hazardous and expensive. The Health and Safety Code of Practice contains recommendations which ensure that the work is correctly undertaken.

New materials

Several studies have been carried out in the United States of America [9–11] which have analysed the importance of materials in construction based on the concept of intensity of use. This is defined as the ratio of the quantity of material consumed by a nation to its gross domestic product. As industry develops, bulk, or conventional materials are used more efficiently and their use declines. This is further influenced by the substitution of more innovative materials which in some measure replace the conventional materials formerly used.

More highly processed, higher value-added materials, account for a larger quantity of construction materials consumed. It is expected that this trend will continue as more closely engineered, or tailored, materials meet technical and reliability criteria. In addition, advanced engineered materials are key components of new technology which contribute to construction capabilities not easily achieved previously. It is necessary to improve quality and productivity to meet increased competitiveness in international markets.

A first stage examination of the intensity of use of traditional bulk materials such as Portland cement and bricks in the economy during the past four decades is shown in Table 4.4 and Figure.4.3.

Significant reduction has taken place in the intensity of use ratios of cement and bricks; usages in 1985 are 62% and 30% respectively, when compared to 1950. The corresponding figures for the USA are 63% and 35% [12]. This suggests that, corresponding to that of the USA, the mode of operation of the construction industry based on the intensity-of-use indicator has changed in 40 years.

New materials that have been introduced, or are being researched at the present time, include the following:

- high performance cement-based materials including fibre-reinforced products
- recycled concrete, asphalt and steel materials
- geosynthetics and geotextile related materials (see Figure 4.4)
- materials to be used in cold regions
- toughened ceramic and ceramic composites

Quality of materials

Before choosing materials to fulfil constructional purposes, it is essential for the user to have a close knowledge of the mechanical and other properties of the material. Many materials, particularly those used for structural purposes, are available to standards which are in accordance with the recommendations of the British Standards Institutions and this provides guidance. Many other materials do not conform with British standards.

In December 1991, the European Commission's Construction Products Directive came into force. This has been already implemented by the United Kingdom. Products must be fit in order to be on the market. The test of fitness is that products have performance in use that is capable of satisfying relevant essential requirements. Appropriate products carry a CE mark and may be sold anywhere in the EU without having to meet any additional technical and

Table 4.4 Intensity of use (IU) of cement and bricks [3]

	1950	1955	1960	1965	1970	1975	1980	1985	1990
Cement (,000 tonnes)	9 752	12 513	13 288	16 704	17 171	16 553	14 805	14 063	17 000
Bricks (millions)	5 921	7 163	7 283	7 868	6 062	5 046	4 562	4 100	3 802
GDP (£bn at 1985 prices)	154	177	202	236	267	295	323	356	417
IU cement	63	71	66	71	64	56	46	40	41
IU bricks	38	40	36	33	23	17	14	12	9

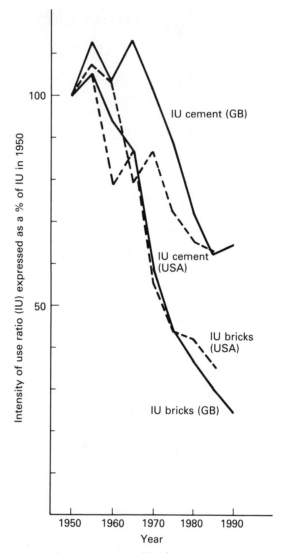

Figure 4.3 *Intensity of use of cement and bricks*

procedural requirements [13]. Products which meet a national stan-dard transposing a harmonized European standard or are the subject of a European Technical Approval (ETA) will be eligible to carry the CE mark. For an interim period, the British Board of Agrément will be the only UK organization to issue ETAs, although product testing may be carried out by approved testing bodies. A European data

Figure 4.4 *How well is this composite material understood by the designer? (a) Geosynthetic material; the grid cells are opened out prior to filling. (b) Grid cells are filled with compacted sand*

bank providing reliable and consistent data on construction materials would be a valuable step in the development of the Single Market.

Users are interested in the properties of materials and have less interest in aspects governing production, particularly when the materials are produced under factory conditions. There is always inherent variability in the properties of materials depending on the method of production and control techniques. Even so, statistical control techniques are applied largely to the crushing strength of concrete and not to most other constructional materials. In addition to testing the end product, a number of inspections are carried out at intermediate stages in the manufacturing process. In this way an unsatisfactory product can be rejected at an early stage. The inspections and measurements provide information about the product and the satisfactory operation of the production process. Manufacturing materials to high standards causes increases in production and inspection costs. There is little evidence to show whether standards adopted in construction are optimal or even the extent to which they are cost-beneficial. However, in terms of the quality control of structural materials, it produces an acceptably low rate of failure due to material deficiencies. Failures that have occurred tend to be more the result of the failure to apply specified control due to human error.

Suppliers and distribution

The bulk of construction materials and products is handled by builders' merchants. Most are members of the Builders Merchants Federation whose registered members control over 2050 outlets with a turnover of some £6500m [14]. Thus, the members of this trade association account for over 40% of materials used by contractors and 30% of all materials used. Builders' merchants have an effective stock-holding of over 20 000 different stock items.

Although some orders of materials are not directly handled by the builders' merchants, an important economic function of the merchants is to build up stocks during periods of small demand, thus smoothing the effect of a variable demand on the materials producers. These stocks can be run down during periods of heavier demand and so avoid hold-ing up construction work due to a scarcity of materials. By combining a number of small orders into a bulk order, builders' merchants can obtain more advantageous prices for purchasers than they could nor-mally obtain. Large contractors are able to obtain advantageous prices by direct purchase of materials in bulk from the producers.

Although the builders' merchant carries on a cash trade, some 80% of sales are to trade accounts. Credit is available for approved accounts and, at any one time, it is estimated that over £600m is owed to the builders' merchants. In this way the merchants appear to act as bankers to the construction industry. They are able to act in this role by means of extended credit from the materials producers which are sometimes their holding companies [15].

Another important function of the builders' merchant is the provi-sion of technical advice on existing and new products, their applica-tions and relevant legislation.

References

[1] NEDO. *How flexible is construction? A study of resources and partici-pants in the construction process.* HMSO, London, 1978

[2] 'Shaking the foundations of the materials industry' in *The Guardian*, 25 January 1992

[3] Central Statistical Office. *Annual Abstract of Statistics.* 1996

[4] Howard, N. 'Energy in balance' in *Building Services*, May 1991

[5] Marsh, R. 'The energy of building' in *The Structural Engineer*, **67**(24), December 1989

[6] Zunz, G.J. 'Matters of concern' in *The Structural Engineer*, **67**(10), 1989

[7] *UK and US Construction Industries: a comparison of design and con-tract procedures.* RICS, 1979

[8] Curwell, S.R. and Marsh, C.G. *Hazardous Building Materials.* Spon, London, 1986

[9] Bever, M.B. and Crandall, R.W. 'Materials economics, policy and management: an overview' in *Encyclopedia of Materials Science and Engineering.* MIT Press, 1986

[10] Larson, E.D., Ross, M.H. and Williams, R.H. 'Beyond the era of materials' in *Scientific American*, **254**(6), 1986

[11] Malenbaum, W. 'Intensity of use of materials' in *Encyclopedia of Materials Science and Engineering*. MIT Press, 1986

[12] Markow, M.J. and Brach, A.M. 'Materials technology in US construction Part 1: Trends in material usage' in *New Horizons in Construction Materials*. Materials Engineering Division of the American Society of Civil Engineers, October 1988

[13] *The single market – construction products*. 1992

[14] Builders' Merchants Federation leaflet. 1991

[15] Colclough, J.R. *The Construction Industry of Great Britain*. Butterworths, London, 1965

5

Technology

The British construction sector faces threats similar to those which have so severely affected automobiles, electronics and shipbuilding. Future skill needs of the construction industries.
Report Prepared for the Employment Department. IPRA Ltd 1991

Design and construction technology

The collection of detailed information and alternative design solutions are developed after receiving the promoter's brief. This is followed by undertaking the detailed design work which pays close attention to the required components and materials to be used in the project but may not consider closely the constructional procedures to be used. Some design teams organized on traditional lines consider this aspect to be the responsibility of the successful contractor, which is not the best way of proceeding with a project. An early involvement of a contractor is advantageous particularly for a project which is novel and complex. Mathematical optimization techniques can be used for the detailed design work but care should be exercised. Optimal solutions are often of limited applicability and do not reflect global cost minimization; in particular, they are not sensitive to construction procedures.

Table 5.1 shows the total cost of constructing an office building in terms of the constituent parts comprising (i) construction, (ii) land and design, (iii) future costs. Considering only the cost of the building, it is clear that the structure is a relatively significant part of the overall cost. When total costs are considered it is not much more than 15%. Modifications to the basic structure are of little significance in terms of relative costs. Although this is true for office buildings, it may also be the case for industrial structures in which expensive machinery and electronics are installed. Nonetheless, initial costs are considered important by the promoter.

Table 5.1 also includes a first stage indication of the total costs for an office building in which the future costs include the salaries of the people working in it. In this case the design and construction costs are 4.0% of the total costs. As a modern office building would include information technology investment and other sophisticated systems requiring maintenance during the life of the building, the design and construction estimate is possibly excessive. A high quality building in which user satisfaction raises productivity by 5% would produce cost advantages of a comparable order of magnitude to the whole of the design and construction costs. This factor has implications for the design of healthy buildings with a high quality internal environment and the avoidance of the sick-building syndrome.

Table 5.1   Relative importance of building costs

Element	Constr.	% costs: office building Capital	Total	Total
Construction costs				
Substructure	6.3	3.8	1.5	0.2
Superstructure	59.5	35.7	14.3	1.4
Internal finishes	7.6	4.6	1.9	0.2
Fittings and furnishings	3.0	1.8	0.7	0.1
Services	17.6	10.5	4.2	0.4
External works	6.0	3.6	1.4	0.1
Land and design				
Land		35.0	14.0	1.4
Design fee		5.0	2.0	0.2
Future costs				
Recurring costs			60.0	6.0
Salaries				90.0

Steyert [1] suggests that a high proportion, i.e. 90%, of the design hours are spent on detailed design work which affects costs by approximately 7% while 10% of design hours are spent on conceptual design work which affects costs by approximately 30%. Hence, detailed design work is subject to the law of diminishing returns, whilst management and financial control of a project play an important role in determining its success. These aspects of construction are highlighted in Chapter 11 which explores the sophistication of management and supervision in the British construction industry.

The construction of structures, particularly large ones, is dependent on a range of well developed-construction techniques. It includes:

- the fabrication and erection of prestressed and reinforced precast concrete members
- the slip-forming of shear walls for large buildings
- the digging of tunnels and erection of precast lining construction
- slip-form pavement construction
- erection of long-span roof construction using reinforced/prestressed concrete shells or cable suspended structures
- reinforced earth structures
- deep foundations comprising piles, rafts or caissons

All these structures require a considerable investment in construction plant. Prefabrication, information technology, the use of new materials and greater mechanization are essential ingredients of technological change leading to an improved performance in the construction industry.

Industrialized building

In some areas of construction, it has long been the practice to fabricate major components of the structure away from the site, often under factory conditions and transport them to the site to be incorporated in the construction. Alternatively, components are fabricated on the site itself but not *in-situ*, i.e. the position that the components will ultimately occupy on the site. Civil engineering has traditionally been the sector of the construction industry that has practised this technique. Many civil engineering projects have relied on the fabrication of beams and other structural components in the factory. The advantage of the controlled environment is that better quality can be achieved when compared to work on site. Site labour is saved but this saving is offset by the extra factory manpower required, nonetheless resulting in overall savings. The preformed components may be made of structural steelwork, reinforced or prestressed concrete. They are accurately cast or fabricated and are transported to site for use in bridges, tunnels and many other structures. Great importance is placed on ensuring that the components are manufactured to an appropriate quality and full account is taken during the design phase of handling, transportation and erection stresses. Precast or prefabricated members are commonly used in composite construction and the use of prefabricated members has improved productivity. There are particular advantages when the site is remote and it is expensive to bring labour to the site to undertake construction work. More recently, prefabrication has found its ultimate goal in the offshore and space programmes where accommodation modules and other components have been constructed to a high quality in order to survive in an aggressive environment before transportation to site and erection.

Although precast concrete offered the convenience of prefabrication it did not, initially, offer the advantages of an industrialized technique. However, the ordered pattern of available sections and the uniformity in making site connections was introduced by the Public Building Frame. The technique was developed by a team of construction specialists from the Ministry of Public Buildings and Works. The characteristics of conformity and simplification provided economy with speed of manufacture and construction; for example, formwork and repetition of procedure was used from one contract to another. The standard components were available for multi-storey buildings and have been used for buildings of over 20 storeys.

The Reema system

Prefabricated industrialized systems can take many forms. The system can be used for both small and large constructions. For example, the Reema construction system [2] used storey-height hollow concrete wall panels for the construction of small houses and blocks of flats up to four storeys in height. External walls of cavity construction and some 200 mm thick, largely unreinforced apart from areas adjacent to openings, were cast from dense concrete. The internal walls incorporated steel reinforcement and were used as loadbearing units. The wall units were cast in the factory and were ready for transportation on the day after casting. Windows were cast into the wall units and were ready-glazed in the factory. Transportation was by road or rail and even in 10 or 15 tonne loads it was economic to transport the units up to 100 miles from the factory. Floor panels could be of hollow construction or constructed from precast concrete beams and timber boards. Standard components weighed approximately 2 tonnes but special components up to 3.5 tonnes could be used. Mobile cranes were required to assemble the buildings. During construction, on a concrete raft laid on a hardcore base, it was necessary to cast corner pillars and horizontal beams around reinforcing bars to form firm connections between the panels and the floor units. This was designed to give the construction an appropriate degree of structural sufficiency. The roofs were conventional pitched roofs or, alternatively, were designed as horizontal and constructed from prefabricated concrete units. Outside finishes to all units were applied in the factory; the internal finishes were skim plaster on the walls or ceilings. The advantages of the Reema system lay in the erection time of the structures when compared with conventional techniques. For example, a pair of semi-detached houses could be erected by a labour force of five men in only 19 days. A comparison of man-hours required for building according to traditional brick construction and the Reema system is given in Table 5.2.

Table 5.2   Comparison of traditional and industrialized structure building time

	Traditional building	Reema building
Man-hours (site)	2 140	1 090
Man-hours (factory)	285	405
Total	2 425	1 495

Not only is a saving of 40% in man-hours claimed but the composition of the labour force is such that a smaller proportion of skilled men is required, i.e. from three skilled men and two unskilled men to one skilled man and two unskilled men.

Prefabricated timber houses have some advantages over concrete and brick-built houses. They have less heat loss; the foundations are lighter and cheaper; erection times are quicker. A small brick house can take nearly 3000 man-hours for completion; in contrast, a comparable timber dwelling takes an average of 1200 man-hours for manufacture and erection of which about 200 hours is used for manufacture of the components.

The Wates system

The Wates system of construction is a panel prefabrication technique used in low and high-rise blocks of flats; blocks in excess of 20 storeys have been constructed. Steam-heated mould plates for the panel units achieve accelerated curing of the concrete panels. The production cycle for the units is 24 hours. Most units are of 3–4 tonnes maximum weight to match conventional foundations. Structural concrete is poured around steel reinforcing bars between the panels. Vertical joints between the panels are made by placing a concrete mix into the facing castellated grooves. It is claimed that the total man-hours are reduced by some 50% and the tradesmens' hours by 60% when compared to conventional construction.

The Bison wall frame system

A flexible system of large panel construction using load bearing walls and precast flooring slabs was developed by Bison, i.e. the Bison wall frame system. It proved economical to use when as few as 24 units were being erected. Blocks of flats with four to eight flats per floor and from 8 to 20 storeys could be constructed using standard units. Each flat used a number of specialized units, for example, single unit bathroom, staircase castings, liftshaft and staircase housings. The standard two-bedroom flat uses a total of 21.5 prefabricated units. The windows are cast into the walls in the factory and can be painted and glazed in the factory. Building services are incorporated in the construction and ducts are cast into the floor and wall slabs to carry services. Underfloor heating is incorporated in the flooring screed. A tower crane is needed on site for erection of the block of flats but no scaffolding is required except when a brickwork outer leaf is used. The Bison factories produced prefabricated units for 8 to 20 dwellings per day. Each flat could be erected by 10 men and it is reported that the saving in erection time over the construction of a traditional building is approximately two-thirds. Hence, a traditional building which takes some 18 months to erect takes only 6 months using the Bison system. Savings claimed in the cost of construction can be as much as 20% in comparison with equivalent traditional methods.

The Larsen/Neilson system

A well known Danish system of prefabrication is that of Larsen and Neilsen. It was adopted by Taylor-Woodrow Anglian Ltd for use in Britain. The building units, which are factory produced, can be used for the erection of small houses, blocks of flats and factories. The wall panels and flooring slabs are cast as large as possible often being of whole room dimensions. The average weight of the units is between 3.5 and 4.5 tonnes. The largest prefabricated components are the bathroom units which weigh approximately 8 tonnes. The panels and slabs are cast in horizontal moulds, the production of each unit taking 24 hours. The units are stored in the open air for at least two weeks before transportation to the site. The cross-walls of the completed building carry the vertical loads down to the conventionally constructed foundations whilst providing, with the longitudinal walls, stability against horizontal wind forces. Mobile cranes erect between two and three flats per eight-hour working day and only half the labour force is required compared with that of traditional construction. For small houses, the actual shell of the house can be erected in a single working day. Even large industrial buildings can be erected quickly.

The impact of industrialized building

Industrialized building has been considered by some to be the panacea to cure the inefficiency in the construction industry in Britain. The characteristics of high productivity are: all fittings that can be prefabricated are made at the fully automated factory; intense specialization is practised at the site with work completely pre-planned; quality is built in during the fabrication and erection processes. During the 1960s in Britain industrialized building received its greatest decade of popularity. During this period it was considered essential to complete as many homes as possible. Government, and opportunistic contractors, used their influence in order to ensure that local authority housing departments recognized that industrialized building was ideal for the rapid production of multi-storey flats. Many public sector housing contracts were let as negotiated contracts or package deals and contractors took responsibility for the design and construction of the blocks of flats. A record 400 000 homes were completed in 1968 with industrialized building playing a significant role. Unfortunately, during this period and subsequent decades, industrialized building contributed to a number of spectacular failures. Poor design and erection procedures emphasized the divisions between design and construction even in the same organization. Also evident was the absence of R&D. Large sums were spent on correcting defects and demolition. For example, the London Borough of Newham contracted Taylor Woodrow Anglian to build nine 60 m high tower blocks at Freemasons Road estate. The Larsen–Neilsen system was used. Two months after completion in 1968 a gas explosion caused the partial collapse of one of the blocks, Ronan Point (see Figure 5.1). It was subsequently discovered that the panels were connected on site with, what proved to be, inadequately designed joints. After the tragedy, remedial strengthening works were undertaken. This was followed by further remedial work 16 years later in 1984 following bowing of the precast panels. A full structural report shortly afterwards effectively condemned the blocks. The blocks were demolished in 1991. During demolition it was found that there had been further lapses in construction methods: mortar

Figure 5.1   *Industrialized building – the experiment that failed!*

contaminated by tin cans, corks and rolled up newspapers; critical hard packed mortar placed to only half the specified depth [3]. Newham Council spent some £3m on demolishing each of the blocks; one is to be refurbished at a cost of £5m. There are 600 such blocks erected in this country. The Association of Metropolitan Authorities calculated in 1984 that £10 000m needed to be spent on correcting design defects in industrialized housing [4].

In 1963 a statement published by the National Joint Council for the Building Industry [5], which is contained in the Working Rule Agreement, recognized that the trend towards industrialization was likely to continue and that the field covered by industrialization would be likely to extend. It was welcomed as a move which can, in the long run, bring benefits to all engaged in the industry in so far as this trend contributed to raising productivity and reducing overall costs. Nonetheless, the National Council sought to protect its members by specifying a number of guiding principles which should be observed whenever and wherever these new forms of building construction were used. These are:

- These new processes and techniques are to be recognized as building operations and should be performed by building trades operatives at the rate of wages and under the working conditions laid down from time to time by the Council.
- The composition of the labour force employed in the work of fixing will necessarily vary according to the material or component concerned. This aspect of site organization will, therefore, need to be approached with a measure of flexibility. As a matter of general principle however, fixing work should be carried out by properly organized teams of building operatives, a broad balance being maintained in the composition of such teams between the numbers of craftsmen and non-craftsmen, as required to meet the demands of the particular operation.
- In order to promote good industrial relations and thereby ensure harmonious working there should, before a new system is introduced or work is started, be joint discussions on anticipated points of difficulty.
- If, during the course of the operation, points of difficulty emerge which cannot be resolved on the site, the issue should be dealt with under the joint machinery for conciliation and the settlement of disputes.

The civil engineering industry has more readily accepted than the building industry that much of the work of construction is undertaken at locations other than the site at which final erection will be established. The Working Rule Agreement published by the Civil Engineering Construction Conciliation Board for Great Britain does not contain a comparable statement to that of the Working Rule Agreement of the Building Industry.

Whereas poor design, supervision and management have contrived to reduce the level of the industrialized building of dwellings from 43% in the 1960s to 2% in 1990 [6], Japan, Germany and France are well equipped in industrialized building technology. In Japan, the complete factory production of customized low-rise housing units is now said to have penetrated 15% of the housing market [7].

The industrial revolution has provided society with techniques for the mass fabrication of most of the artefacts required by large populations. This work is undertaken in factories in which the environment is controlled, there is a close commitment to health and safety and the quality of the product is maintained and assured. Often robots carry out most of the repetitive work and have protected the labour force from drudgery and heavy work. It is strange that attempts to move the British construction industry more emphatically from handicrafts technology to a more modern, up-to-date industry have failed.

Maintenance and performance monitoring

Until fairly recently the construction industry, unlike the engineering or chemical industries, paid relatively little attention to the performance monitoring of its artefacts. Although it was recognized that the deterioration of a structure commenced during construction and continued with time, even maintenance was undertaken only when structures showed signs of distress. However, the advantages of maintenance and performance monitoring have been recognized [8] and greater attention is being paid to the monitoring of building structures using surveying and photogrammetric techniques, dynamic testing and automatic monitoring [9]. To counter rapid deterioration of buildings it is essential that the contribution of the designer, builder and building surveyor is properly undertaken. Yet there is no basis for allocating the balance of expenditure on design, checking, supervision and inspection. There are no cost–benefit systems for error control. Standards are set for the appropriate levels of materials quality and workmanship without a clear understanding of how this impacts on the safety of the structure. Performance monitoring confirms that the behaviour of the structure is in accordance with the predicted behaviour; gives knowledge of any discrepancy in performance and gives an opportunity for corrective action; improves the body of knowledge that is required to undertake design with confidence thereby improving future structures [10]. Offshore structures and construction activities have, in particular, been subject to close performance monitoring as the detection and repair of fatigue damage is essential. There are considerable uncertainties in the prediction of fatigue lives and scope for the development of a reliable theory to indicate the relative risk levels of fluctuating loading. The development of performance monitoring techniques has created an alliance between the construction industry and instrumentation engineers. Sophisticated, reliable and robust instrumentation has been developed to provide a volume of data on structural behaviour. Instrumentation includes strain gauges, pressure sensors and accelerometers. Nearly 100 different instruments were used on the Stratford B Condeep Platform to monitor installation; 116 are used for performance monitoring during use. There is considerable redundancy in instrumentation, but the failure rate has been less than 2% even in the aggressive environment of the North Sea.

Examples of other constructional work being monitored [11] include:

- office blocks in which an assessment is being made of the floor and column performance prior to refurbishment
- asphalt membranes in dam construction
- buildings close to new tunnelling work and other new developments in congested city sites
- 30 m high diaphragm walls supporting deep excavations in clay

It is expected that there will be considerable improvements in performance monitoring techniques as structures increase in size and complexity.

Intelligent buildings

Business in the 1990s often requires buildings to be 'intelligent' or, in United States terminology, 'smart'. An intelligent building is one that satisfies the requirements of its occupants. This might include flexibility which caters for different work groups with varying needs at locations that may be subject to frequent change. The latter factor is often referred to as the 'churn rate', i.e. the number of people moving location within a building each year; this can approach or even exceed 100% [12]. The operation of business is dependent on massive information technology investment which facilitates information exchange within buildings and between buildings. This is more than enhanced telecommunications; these services include word processing, electronic mail, video conferencing and networks which access a wide range of business databases. Electronics contribute to the handling of the environment, security, the regulation of access including lift movements, energy and lighting management and operation. Futuristic concepts include structures that can respond in a more sophisticated way to the changes in the external environment using insulated glass with variable opacity, i.e. 'smart' glass, materials with variable thermal capacity, self-cleaning and maintenance services.

Many existing buildings cannot cope with these new demands. For example, offices built in the 1960s and 1970s, even in locations such as the City of London where building intelligence is of prime importance, may not be able to be brought up to the information technology specifications required or to allow radical organizational change. This is because these buildings have physical limitations which restrict vertical ducting and ceiling level distribution of cables or air trunking. The lack of rapid space reorganization is a further limitation. An example of a modern intelligent building is that of the Lloyds' building in the City of London. As the services are on the outside of the building, the floors are unaffected by service ducting. This, in turn, increases office layout flexibility and keeps the costs involved in the relocation of activities to a minimum.

Building change needs management. The more intelligent the building, the more difficult it is to manage. To reorganize heating, cooling, lighting, security and other facilities requires more sophisticated systems and highly trained personnel to carry out their operation. To provide advice and guidance to these personnel an expert system can be used [13]. Indeed, there is a view that an intelligent building is one with an expert system incorporated to facilitate building management. The Building Research Establishment has developed an expert system (BREXBAS) which monitors sensor data from remote systems, applies its knowledge and reasoning capabilities to interpret the information and generates advice for the user. Greater control is gained over increasingly expensive building space enabling it to be used more effectively and efficiently.

Many office buildings are built as speculative ventures. It is, therefore, difficult to know the precise demands of the occupier. The developer must design the building with the facilities to develop intelligence to the degree required by the operator incorporated in the building fabric. This includes ducting, cabling routes, adequate floor to ceiling height and flexibility in building services. Schemes such as Broadgate, Canary Wharf and new style business parks on city fringes rely on the developers' knowledge of user requirements. Sectoral studies of user requirements extend present knowledge on the variation of requirements between, for example, banking and insurance. Ultimately, it is a matter of optimizing building performance during the entire life of the building against a range of potential requirements.

Robotics in construction

Construction work requires the execution of dangerous and demanding tasks in an environment which can be dark, dirty and, in the longer term, injurious to health. In order to relieve construction workers from the rigours and hazards of harsh operational environments there is potential for the development and application of robotic devices. Such devices have been used successfully in the nuclear and steel industries but construction has not fully benefited from these technological advances. With these applications, it is expected that there will be the additional benefits of increased productivity and the improvement of quality. These advantages have been seen more clearly in Japan [14] where the development of construction robots is most advanced (Figure 5.2). In 1988, Brown [15] cited 89 examples of construction robots of which 74 were Japanese. The hardware had been developed for 71 of the Japanese robots and in nine other examples. Each of the top six construction companies in Japan has produced working prototype robots, although whether they are yet fully cost-effective for day-to-day site operations, even in a land where site labour is highly paid, is questionable. Nonetheless, they offer considerable time savings when their output is compared with that of a skilled operative.

Attention is not presently focused on a co-ordinated approach to construction integration and automation. Rather, separate robots capable of fulfilling a variety of different functions have been developed. These include robots for:

- erecting steel structures
- finishing concrete floors
- site welding
- painting

A frequent objection to the use of robots is that the construction site is not orderly, as is a manufacturing operation, but constantly changing, with the multiplicity of craft operations that are carried out not being suited to replacement by robotic devices. Notwithstanding these objections, there have already been a number of successful robotic devices used in construction and there is little doubt that these developments will continue even though, at this stage, it is not possible to define what their developed forms may be. Presently robotic forms fall within three classes each of which is identified by the control procedures available to the robot and its relationship to the human operator or supervisor. Whittaker [14] has defined these three classes in the following way:

1 Teleoperated robots: these include machines where all planning, perception and manipulation is controlled by humans.
2 Programmed robots: these perform predictable, invariant tasks according to pre-programmed instructions.
3 Cognitive robots: these sense, model, plan and act to achieve goals without intervention by human supervisors.

Although classes 1 and 2 are the most common, there are examples of class 3 which are relevant to constructional environments. All three classes will continue to find relevance in construction activities. Hybrid forms are also available.

Figure 5.2   *Construction robots used by the Japanese. (a) Concrete slab finishing robot (Kajima corporation). (b) Underwater rubble levelling robot (Penta-Ocean Construction Co. Ltd)*

Shimizu Mighty Jack; an example of a teleoperated robot [16]

The erection of steel beams is a dangerous task. The Mighty Jack manipulator lifts two or three steel beams and sets them in the correct position. Before setting the beams the manipulator is lifted into position and grasps the tops of the columns to which the beams are connected. Cranes are not required during the actual beam setting and can be used for other operations elsewhere on the site. The connection of the beams is undertaken manually. The robot takes some 25 minutes to set six beams. Using traditional procedures takes 40 minutes. In addition to increased productivity, it is judged that the work is carried out more safely using the Mighty Jack manipulator.

The Mighty Jack manipulator is an immature form of robot. The man–machine interface is not ideal as the optimal process is lost through the coupling procedures. However, it is expected that the versatility of the Mighty Jack will be improved through R&D.

Another example of a teleoperated robot is the Ohbayashi–Gumi concrete placer. This robot eliminates heavy, dirty work. It carries out its task to a good quality finish and is reported to reduce concrete placing labour costs.

Mobile robot for concrete slab finishing Mark I (Kajima Corporation); an example of a programmed robot [17, 18]

The concrete slab finishing robot, Mark I, carries out the several finishing stages and the final floating off required for a high quality finish. It relieves the operative of physically demanding work over long periods of time, sometimes throughout the night. The robot is programmable in order to cater for slab shapes of a more complex nature or to cope with columns that may be present. A specially trained engineer is required to input the initial data, for example, locations and dimensions of columns. Quality control tests showed that the robot finished the slabs with variations in horizontality of the surface of only plus/minus 1.94 mm over a distance of 3 m; there was a maximum variance of plus/minus 3.50 mm. These figures are well within the tolerance limits of plus/minus 7 mm over 3 m, and are comparable with the results obtained by a skilled operative.

Since developing the Mark I robot, there have been substantial improvements to develop the more sophisticated Mark II. The Mark II is considerably lighter and needs less power. In addition, a smaller computer is used, mounted in the robot itself, together with a lightweight gyrocompass. Little input information is required as a touch-sensitive bumper allows the robot to navigate around any intervening object such as a column. It is also equipped with radio-control that permits manual operation if required. The Mark II robot obtains a superior slab finish to that of the Mark I and it has a floor slab finishing capability of 500 m^2 per hour. This considerably reduces the number of site operatives required. It is expected that there will be a further improvement to performance characteristics and cost-effectiveness.

Other examples of a programmed robot include the Shimizu Insulation Spray Robot and the Kajima Reinforcing Bar Arranging Robot.

GENEREX, a general robotic excavator; a speculative example of a cognitive robot [19]

The GENEREX is based on an earlier prototype of a Robotic Excavator (REX) which integrated sensing, modelling, planning, simulation and action specifically to unearth buried gas utility piping. REX is comprised of hydraulic actuation hardware in the form of an arm mounted on a utility truck. Air-jet tooling is deployed and controlled by means of a sonar sensor which collects range data to construct a terrain model of the excavation site.

REX is reported to reduce the excavation hazard associated with the explosive gases, decrease the operation costs and increase the productivity when used in the gas utility industry. Nonetheless, a critical evaluation of REX reveals a number of shortcomings linked to tooling, manipulation, control, sensing, modelling and strategy. A second-generation excavator, GENEREX, has been postulated which is an improvement on REX and indicates the way forward for a cognitive robot system. There is provision for the accommodation of human interactions at a number of levels ranging from observation to full teleoperational control which may be necessary in highly unstructured environments.

Robots under development

Prototype robots are currently under development in several countries in the world. Britain is adapting a robot presently used to inspect nuclear pressure vessels, for use in construction applications. Wall climbing robots for inspection work, tunnelling systems and underwater robots are being closely studied. Systems operating in mainland Europe include a wall-tiling robot and a robot to paint buildings [11]. Although construction robots are at an early stage of development, it seems clear that their introduction can lead to potential cost savings and improvements [20]. The safety of site operatives will be significantly improved. When robot construction systems are more fully introduced, the rewards will be considerable. This has been recognized in countries such as Japan and the USA where a considerable R&D effort is taking place.

The International Association of Automation and Robotics in Construction was formed in 1991 with the objective of promoting the application and benefits of automation and robotics in construction operations and maintenance. The range of interests include the built environment, offshore structures and constructions in space.

References

[1] Steyert, R.D. *The economics of high-rise apartment buildings of alternative design configuration.* American Society of Civil Engineers, New York, 1972

[2] Diamant, R.M.E. *Industrialized Building.* Iliffe Books Ltd, London, 1968

[3] 'Point of contention' in *New Civil Engineer*, September 1991

[4] *Construction News*, November 1984

[5] Working Rule Agreement. National Joint Council for the Building Industry. 1991

[6] Department of the Environment. *Housing and Construction Statistics.* HMSO, London

[7] *Construction research needs in Europe.* European Network of Building Research Institutes, Luxembourg, 1991

[8] *Planned Building Maintenance. A Guidance Note.* RICS. 1990

[9] Moore, J.F.A. *Monitoring Building Structures.* Blackie, London, 1992

[10] 'Field measurements in Geotechnics'. *Proceedings of the International Symposium.* Zurich, September 1983

[11] *News of construction research.* Building Research Establishment. December 1991

[12] 'Smarter than the average building' in *Management Today*, December 1990

[13] 'Building an IQ. Computer Aided Construction' in *New Civil Engineer Supplement*, January 1990

[14] Whittaker, W.L. 'Construction Robotics: A Perspective. CAD and Robotics in Architecture and Construction'. *Proceedings of the Joint International Conference.* Marseilles, June 1986

[15] Brown, M.A. 'The application of robotics and advanced automation to the construction industry'. *Occasional Paper No. 38.* The Chartered Institute of Building, 1989

[16] Ueno, T., Maeda, J., Yoshida, T. and Suzuki, S. 'Construction robots for site automation. CAD and Robotics in Architecture and Construction'. *Proceedings of the Joint International Conference.* Marseilles, June 1986

[17] Tanaka, N., Saito, M., Arai, K., Banno, K., Ochi, T. and Kikuchi, S. 'The development of the Mark II mobile robot for concrete slab finishing. CAD and Robotics in Architecture and Construction'. *Proceedings of the Joint International Conference.* Marseilles, June 1986

[18] Arai, K. *Construction robots that aid in advanced construction.* Kajima Corporation, 1990

[19] Whittaker, W.L. and Motazed, B. 'Evolution of a Robotic Excavator. CAD and Robotics in Architecture and Construction'. *Proceedings of the Joint International Conference.* Marseilles, June 1986

[20] Skibniewski, M., Derrington, P. and Hendrickson, C. 'Cost and design impact of robotic construction finishing work. CAD and Robotics in Architecture and Construction'. *Proceedings of the Joint International Conference.* Marseilles, June 1986.

6

Construction firms

The term fluidity has already been used in relation to the construction industry, but perhaps versatility is better or even diversification.

J R Colclough, 1964

Numbers and types of firms

At the start of the 1990s there were estimated to be almost 210 000 construction firms operating in Britain. These had fallen by 15 000 by the mid-1990s. These firms range from a single tradesman, or sole proprietor, perhaps earning a living from what is known in the industry as jobbing building, to the huge conglomerates of businesses employing several thousands in their workforce. Construction firms start up, grow, merge with other firms, break up and sometimes die. Some thriving, long-established firms of just a few years ago have since ceased trading or been taken over by other companies. Construction firms can be grouped and organized in many different ways according to their main sphere of activities, their location, the number of their employees, the size of their annual turnover, their capital resources, or in several other ways, and in any combination or permutation that might be thought desirable. There are difficulties of classification, due in part to the individuality of the different firms, and the diverse nature of some of their activities and descriptions that are sometimes not evidenced by their practices.

Table 6.1 and Figure 6.1 indicate that in 1970 there were an estimated 70 000 construction firms. Of these approximately 210 (0.28%) had a workforce of over 600 employees. Twenty years later in 1990, the total number of firms had increased threefold to almost 210 000,

Table 6.1 Number of construction firms [1]

Total employment	1970	1980	1985	1986	1987	1988	1989	1990	1991	1992	1993	1994
1	20 355	36 549	72 896	76 946	79 354	83 484	94 218	101 223	103 169	94 452	93 585	97 141
2–3 }	33 118	34 522	54 405	54 223	54 712	57 878	67 189	71 498	70 452	68 486	64 438	65 188
4–7 }		20 586	24 171	24 455	24 838	25 639	24 984	23 403	21 664	30 395	26 072	22 145
8–13	7 946	10 052	7 164	7 067	7 074	6 156	5 869	5 362	4 981	5 240	4 630	4 221
14–24	5 358	5 849	4 582	4 520	4 485	4 306	4 212	3 935	3 429	3 574	3 129	2 881
25–34	1 982	2 002	1 519	1 394	1 507	1 467	1 478	1 420	1 186	1 146	1 066	956
35–59	2 062	1 985	1 480	1 502	1 520	1 471	1 458	1 305	1 100	1 148	1 098	1 008
60–79	720	592	441	448	456	451	450	442	382	361	294	325
80–114	616	484	409	369	393	406	421	432	372	317	283	262
115–299	820	663	512	501	507	510	530	507	431	387	330	356
300–599	233	208	141	130	143	138	153	150	137	103	96	92
600–1199	132	92	66	71	71	70	66	69	58	59	53	50
1200 and over	78	48	39	34	35	42	48	47	39	36	33	32
Totals	73 420	113 632	116 186	144 395	160 596	169 999	167 825	171 660	207 400	205 704	195 107	194 657

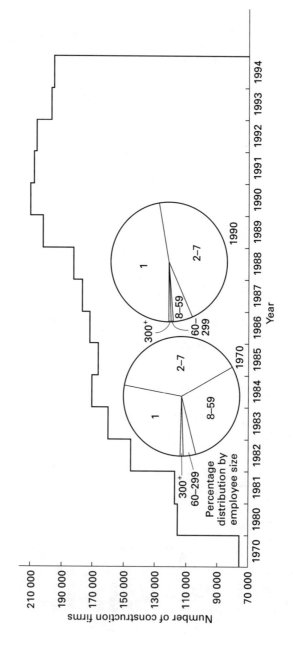

Figure 6.1   *Numbers of construction firms [1]*

but of these only 110 (0.05%) had more than 600 employees. The larger firms of over 1200 employees had also declined in number over the two decades. This continues to be the trend. There were 78 large firms in 1970, reducing to 48 (1980), 47 (1990) and 32 in 1994. These figures partially reflect the recession in the industry, mergers and down sizing amongst construction firms. The recession in the early 1990s is probably the single factor that has resulted in a reduction in the numbers of firms across all size classifications of firms. By the mid 1990s over 15 000 construction firms had disappeared since the peak year of 1990. This represented a decline of over 7%. Conversely there had been a dramatic increase in the number of small firms. These figures help to explain the rapid and widespread increase in the subcontracting business that occurred over this period of time. The one-man firm increased by a factor of five from about 20 000 in 1970 to over 100 000 by 1990. In 1970 these firms represented 28% of the total number of construction firms. By 1990 they represented almost 50% (48%). They had grown not just by number but also by proportion. The growth in the number of firms has shown a steady trend throughout this period of time. The firms of below eight employees have grown in number, whilst the firms above this size have in general tended to shrink in number. This further emphasizes the shift towards smaller firms of subcontractors. Of the extra 135 000 firms which came into existence between 1970 and 1990, over 80 000 represented one-man businesses. Construction activity today involves a multiplicity of different firms.

Of the total number of construction firms (210 000) in 1990, almost 40% described themselves as general builders (Table 6.2). Proportionately, this represents a small decline over the two decades in respect of the total number of firms in business. The combined numbers of building and civil engineering contractors in 1970 was 35 314 and in 1990 it was 87 496. This represents a 47% increase, but the figures are partially masked by size, and the descriptive aspirations some firms may employ in order to attempt to secure what they believe may be the more lucrative work. These general contractors represented 31% of the total number of firms in 1970 and 42% in 1990. The major subcontracting firms, i.e. plumbers, carpenters and joiners, glaziers, etc. increased their numbers from 27 803 in 1970 to 68 433 in 1990. These represented 25% and 33% of the total number of firms respectively, indicating a fairly large growth in this type of firm over the two decades. There are no brickwork firms recorded, even though these do exist in the construction industry. It is likely that they are included under the description of general builder. Interestingly, the number of carpentry and joinery firms is lower than painting, plumb-

Table 6.2　Classification of firms by trade of firm [1]

	1970	1985	1986	1987	1988	1989	1990	1991	1992	1993	1994
General builders	31 546	67 475	68 320	68 947	70 476	77 222	78 981	76 991	74 393	70 765	69 160
Building and civils	2 201	3 623	3 777	3 917	4 075	4 522	4 773	4 772	6 180	6 264	6 845
Civil engineers	1 567	2 662	2 847	2 946	3 108	3 468	3 742	3 760	4 312	4 070	4 182
Plumbers	6 718	14 934	15 030	15 207	15 570	16 774	17 046	16 752	14 647	13 880	13 181
Carpenters and joiners	4 572	10 949	11 337	11 741	12 408	14 033	15 244	15 499	14 199	13 302	12 614
Painters	12 268	14 662	14 719	14 898	15 053	16 118	16 511	16 142	10 788	9 774	8 974
Roofers	1 349	5 818	6	6 145	6 487	7 294	7 767	7 722	7 524	6 891	6 470
Plasterers	2 615	4 019	4 128	4 137	4 201	4 567	4 834	4 761	3 893	3 549	3 160
Glaziers	281	4 387	4 640	4 871	5 396	6 231	6 531	6 435	7 001	6 599	6 918
Demolition contractors	284	559	608	634	616	656	667	635	717	708	685
Scaffolders	96	966	1 028	1 081	1 179	1 401	1 524	1 597	1 779	1 645	1 733
Reinf. conc. spec.	240	515	551	597	632	733	910	999	859	729	637
Heating and Ventilating	2 211	8 461	8 614	8 751	8 951	9 485	9 624	9 375	9 774	9 355	9 136
Electrical	4 095	15 449	15 993	16 402	17 139	19 215	20 732	21 020	21 780	20 589	21 004
Asphalt	303	856	887	913	945	1 036	1 080	1 081	1 163	1 071	1 077
Plant hire	1 475	3 664	3 723	3 819	4 009	4 473	4 628	4 579	5 621	5 567	5 940
Flooring	438	1 400	1 451	1 495	1 611	1 841	1 999	2 033	2 387	2 248	2 320
Constr. engineers	363	1 560	1 694	1 756	1 919	2 286	2 592	2 640	2 713	2 375	2 168
Insulating spec.	84	1 308	1 341	1 379	1 375	1 399	1 376	1 281	1 265	1 147	1 131
Suspended ceilings	76	842	922	1 011	1 137	1 381	1 554	1 659	1 757	1 597	1 509
Floor/wall tiling	298	1 167	1 222	1 259	1 325	1 516	1 657	1 675	1 607	1 492	1 430
Miscellaneous	350	2 549	2 819	3 189	4 136	5 425	6 003	5 992	11 345	11 490	14 383
Totals	73 430	167 825	171 660	175 095	182 018	201 076	209 793	207 400	205 704	195 107	194 657

ing and electrical contractors, although in terms of the numbers of people trained in these crafts there are considerably more carpenters and joiners. There may be several explanations for this. For example, many more carpenters and joiners, than the other trades, move upwards into jobs of site management or work under the description of 'joiners and builders' and thus classify themselves as general builders. The costs of setting up their businesses along with the expensive woodworking equipment and the need for a workshop of some kind also makes their firms either more of a partnership or a general building organization. Proportionately the number of firms in the demolition business declined from 0.4% to 0.3% of the total number even though the actual number of firms increased by almost 40%. All the specialist subcontracting firms showed a considerable growth in their numbers. Whilst the general contractors under the different classifications predominate in the larger organizations, amongst the smaller firms all different specialisms are represented.

Table 6.3 and Figure 6.2 indicate the amount of work undertaken by the different sizes of firms. The total amounts of £55.31bn (1990) and £49.44bn (1994) are based upon the output by contractors, including estimates of unrecorded output by small firms and self-employed workers and labour departments. It is classified in accordance with the 1980 Standard Industrial Classification. The figures indicate the peak of construction activity in 1990, which was then followed by the recession in the industry of the early 1990s. Between 1990 and 1992 an almost £8bn (14%) reduction in the workload based upon current prices was experienced by the industry. The effects of the recession were much greater, since much of the work carried out in the early 1990s was done so at cost and the figures are current prices allowing for inflation.

Table 6.3   Amount of work undertaken by contractors [7]

Firms size (employees)	New work		Repairs and maintenance				Totals	
			Housing		Non-housing			
	1990 bn	1994 bn	1990 bn	1994 bn	1990 bn	1994 bn	1990 bn	1994 bn
1–7 employees	4.28	4.64	8.39	8.78	2.80	3.59	15.47	17.01
8–114	8.64	7.93	3.71	3.50	4.46	4.23	16.81	15.66
115–1199	10.40	8.25	1.39	0.95	2.51	2.31	14.30	11.51
1200 and over	7.44	4.27	0.35	0.54	0.94	0.45	8.73	5.26
Totals	30.76	25.09	13.84	13.77	10.71	10.58	55.31	49.44

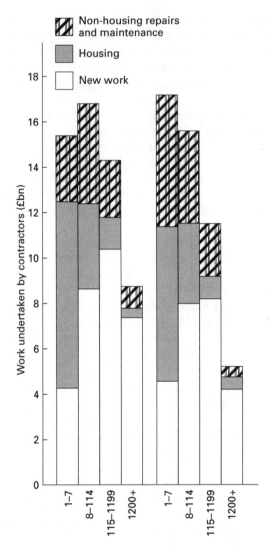

Figure 6.2   *Amount of work undertaken by contractors (1990 and 1994) [1]*

Of the £55bn of contractor's output in 1990, £31bn (56%) was defined as new work. By 1994 the amount of new work had reduced to £25bn and as a proportion of the total work carried out (51%). The amounts spent on repairs and maintenance over this period stayed the same, although in real terms the amounts declined.

The smaller firms did marginally better than the larger firms but only amongst the very small firms did output actually increase. In 1990, it was worth £15.47bn (28%) and by 1994 these figures had increased to £17.01bn (34%). The larger firms fared much worse. Many reduced the numbers of their employees considerably and the actual number of firms in this classification reduced by one-third (Table 6.1). Their output fell from almost £9bn to just over £5bn, down by a massive 40%. In 1990 the larger firms share of the output was 16% but by 1994 this had fallen to 11%.

Table 6.4 shows the percentages of the different types of work that is undertaken by the different size of contractor groupings for 1990 and 1994.

Table 6.4   Type of work undertaken by contractors 1990 and 1994 [7]

Firms size (employees)	New work		Repairs and maintenance			
			Housing		Non-housing	
	1990 %	1994 %	1990 %	1994 %	1990 %	1994 %
1–7 employees	13.9	18.5	60.6	63.8	26.1	33.9
8–114	28.1	31.6	26.8	25.4	41.7	40.0
115–1199	33.8	32.9	10.1	6.9	23.4	21.8
1200 and over	24.2	17.0	2.5	3.9	8.8	4.3

Whilst in 1990 almost a quarter of all new work was carried out by the largest firms, i.e. those with over 1200 employees, this had fallen to only 17% by 1994 (Table 6.4). The smaller firms predominate on housing repairs and maintenance, but the firms in the next size category continue to take the largest share of the non-housing repairs and maintenance. The very large firms are hardly involved in these activities at all. These figures clearly support the age of the subcontractor or trade specialist. Even on large projects whilst the bigger firms will take overall responsibility, much of the actual building work is subcontracted to the smaller firms and organizations.

Table 6.5 shows a comparison of total employment by contractors for the three years of 1985, 1990 and 1994. This table shows clearly the massive reduction in employment levels between 1990 and 1994 of almost a quarter of a million. Some of these transferred to self-employment within the construction industry, but many left the construction industry on a permanent basis. Over this period the larger contractors reduced their employees by over 40%. All sizes of firms showed a reduction in the numbers of their employees. Proportionately from 1985 there has been an increase in the percentage of employees working for the smaller firms.

Table 6.5   Total employment by contractors [7]

Firms size (employees)	1985		1990		1994	
	,000	%	,000	%	,000	%
1–7 employees	296.6	32	367.2	37	322.6	43
8–114	332.1	35	295.3	30	215.4	28
115–1199	203.8	21	208.9	21	143.1	19
1200 and over	108.6	12	125.3	12	72.2	10
Totals	941.1	100	996.7	100	753.3	100

Table 6.6 shows the regional distribution of construction firms between 1970 and 1994, ignoring the fact that firms often do not confine themselves to specific geographical regions. Approximately 87% of firms in 1970 were in England. This increased to 89% by 1980 and then levelled off. In 1970, 34% of firms were located in the south east. This proportion declined to 32% in 1980 but had further increased to 37% by 1990. The area that showed the biggest increase in the number of firms over the two decades was, not surprisingly, East Anglia with four times as many firms in 1990 than in 1970. This increase was reflected in the population and economic growth of that area over those decades. The increase in the regional number of firms is a limited indicator, since some of these firms will undertake work outside of their regional boundaries. It can be assumed that the regional changes are generally in line with the national patterns in terms of size. The regional pattern mirrors the general reduction in the numbers of construction firms. The region of East Anglia, that increased its number of firms by the largest amount between 1970 and 1990, also showed the largest decline in the early 1990s. However, it still represents the largest number of firms size growth over the quarter century.

British contractors

It is difficult to make comparisons across British firms because of the varying commercial interests that the different contractors have in materials producers, quarries and product manufacture [2]. Some of the larger firms have also diversified into activities beyond the remit of the construction industry. Using a strict definition criteria, Tarmac's building materials and quarry products turnover and profits, should be excluded. This would mean that Trafalgar House would be the larger firm. Other companies also have non-contracting busi-

Table 6.6   Regional distribution of Contractors [1]

	1970	1980	1990	1994	Percentage increase 1970–1994
North	3 390	4 696	7 544	7 333	122
Yorkshire & Humberside	6 715	9 627	15 772	15 290	128
East Midlands	4 916	8 746	14 182	13 877	182
East Anglia	2 461	4 852	10 376	8 953	263
South East	24 753	37 037	77 780	68 665	177
South West	6 737	11 382	23 928	20 032	197
West Midlands	6 893	10 935	18 471	17 740	157
North West	8 376	13 839	19 393	17 800	113
England	64 241	101 114	187 446	169 690	164
Wales	3 729	5 533	9 612	10 269	175
Scotland	5 450	6 985	12 735	14 698	170
Totals	73 420	113 632	209 793	194 657	165

nesses and it would be difficult to remove these. In addition, the narrowest definition of contracting should exclude house building, an activity in which many of the large firms now have an active share. Such a definition would become less representative of the construction industry as a whole. During the past 20 years there have been many amalgamations and mergers, resulting in firms of conglomerates who now dominate the construction business in Britain. Many of the firms which contribute to these large groups are household names, even to those who have no connection with the industry. In a few cases an identity has been lost in favour of a new group holding company name which identifies the different constituent groups as specialist divisions within the whole. The loss of even some of these names has removed a little bit of folklore from the industry. Table 6.7 provides a list of the top ten contractors in Britain [3].

Many of the larger construction firms are now a part of parent companies which have multi-interests across a wide, diverse section of different industries. Trafalgar House, for example, which includes Trollope and Colls the construction firm and engineering companies such as Davy and John Brown, also have non-construction firms within their conglomerate organization. For example, Trafalgar House owns the Ritz Hotel, which is currently estimated to be valued at £150m, and Cunard which includes the world's most famous

Table 6.7 The ten top contractors of Great Britain, ranked by turnover [2] [5]

Rank	Company	Turnover (1994) (£m)	Profit (1994) (£m)	No of Employees	European Ranking
1	Tarmac	2 515	107	15 350	11
2	AMEC	1 966	29	5 804	18
3	Wimpey	1 726	45	7 265	22
4	Balfour Beatty	1 549	50	16 115	24
5	Bovis	1 524	24	3 511	25
6	Mowlem	1 358	5	12 184	26
7	Laing	1 174	24	6 425	27
8	Taylor Woodrow	1 149	51	5 129	35
9	Trafalgar House	1 148	25	9 180	36
10	Costain	1 007	—	11 000	41

ocean-going liner the *Queen Elizabeth II* (QE2). The engineering and construction division, however, accounted for 80% of the group's £3.45bn turnover in 1989/90. In 1990 Trafalgar House had a huge cargo shipping interest which it sold to P&O. As a company the Trafalgar House group have indicated that their days of the conglomerate are probably over. It considers that more can be gained from specialization. In 1991 the group acquired the construction company, A. Monk PLC, with its £248m turnover. The engineering and construction division of Trafalgar House is one of the largest in the world in terms of construction size, but a way behind the USA's Bechtel and Fluor, currently the world leader in construction and engineering projects.

Comparing the 1994 league table (Table 6.7) with that of previous years, the names remain the same although their respective positions have changed due to many different factors. As long ago as 1989, it was predicted by some industry commentators that the number of major contractors in Great Britain would reduce to about four or five major firms. The present-day construction industry has similarities to the car industry of the 1950s. By the end of the recession, not only will it be much smaller, it will also be foreign owned. Razor thin margins are forcing firms who choose to stay in contracting to become very different companies. Most have restructured in some way, becoming more specialist and shedding hundreds of staff. Some are heeding the words of Latham [7] by introducing new styles of management and partnering with major clients.

Tarmac has swapped its housing division for Wimpey's contracting and minerals. This now makes Wimpey the largest house builder in Great Britain. AMEC has for the time being fought off an hostile take-over bid from the Norwegian shipbuilder Kvaerner. Trafalgar House has become a subsidiary of this foreign company and has sold its Ideal Homes to the house builder, Persimmon. Laing is seeking to concentrate on the higher margins of negotiated work and private finance initiative schemes. It has closed its small works division and will not bid for work below £2.5m. Taylor Woodrow has been relying upon overseas contracting to compensate it for poor performance in the home market. Costain has been rescued by a Malaysian conglomerate.

Construction Employers Associations

The construction industry has typically been seen as two sectors; building and civil engineering. This distinction is also seen at employers' level with the separate Building Employers Confederation (BEC) and the Federation of Civil Engineering Contractors (FCEC). A third group, the Export Group for the Constructional Industries, includes those contractors with extensive overseas interests. These organizations seek to represent the contractors' interests at national level and to influence government policy for the benefit of their members. Most of the largest 20 firms belong to all three organizations. The BEC has a membership of 8000; this was over 10 000 prior to the recession of the early 1990s. The number of firms in the FCEC is much smaller at 300 (700 previously), although the typical size of firm is much larger. In addition, there are other subsidiary groupings of firms such as the House Builders Federation. The fragmentation is seen as reducing the single voice of the contractors. Many of the activities which are undertaken by the different organizations are replicated and add to the costs of administration. There has also, in recent years, been an increased level of co-operation between the different bodies on matters such as European harmonization and health and safety. These activities are seen as steps towards an eventual merger representing all firms with subsidiary committees for special interest groups. Many of the contractors believe that the construction industry's poor image, educational framework, lobbying strength and relationships with clients and the professions will be improved only through such a united single voice. At the time of going to press FCEC is likely to be disbanded.

The Major Contractor's Group (MCG)

The MCG is a group comprising the largest contracting companies and includes about 75 firms. The group has its own code of practice to ensure high standards of reliability and efficiency. Its members must amongst other things:

- develop and maintain the good name of the member firm by conducting its activities in a business-like manner
- carry out work with proper consideration for and minimum disturbance to the general public
- observe recognized standards of good tendering practice
- use every endeavour within the limits of the contract conditions to:
  - complete contracts on time and within cost limits
  - fulfil all their obligations under the contract
  - help to promote the client's understanding of the contract
- co-operate with and form good relations with the client's consultants
- observe the National Working Rule Agreement and support the National Joint Council for the Building Industry
- act in accordance with good business practice when dealing with other companies
- maintain good industrial relations
- have proper regard for recognized standards of safety and welfare and set an example in accident prevention
- keep pace with technology and the development of modern techniques and methods
- provide adequate training and the development of employees to ensure the provision of skills and expertise to meet the needs of a progressive company
- keep the public informed of the nature of the contracting operations

Within all organizations there are tensions, particularly where a group do not believe that they receive adequate representation. As the major 20–25 firms have grown, and the issues which they feel need to be addressed are now different from the other firms, their focus is towards the international business environment.

Chartered Building Company Scheme (CBCS)

In 1993 the Chartered Institute of Building introduced its Chartered Building Company Scheme. One of its aims is to attempt to improve the generally accepted poor image of the construction industry. Attempts to improve this image must be welcomed. The scheme is not a guarantee of workmanship or a warranty against defects but that the company is run by professionally qualified people. The scheme lays down few rules, but in order for a company to use this title at least 75% of its directors must be professionally qualified and half of these must be members of the CIOB. The argument for this is that those companies which are professionally run rarely let complaints turn into disputes. Companies are expected to pay between £250 and £1500 to enrol, depending upon their turnover. An independent board within the CIOB will administer the scheme. Any unresolved disputes which arise will be referred to this board which will, if necessary, appoint an arbitrator. A code of conduct will be established which will ensure that companies discharge their duties to their clients and employees and employ competent specialist contractors. A CBCS firm should strive to offer clients a quality service which provides them with value for money.

European contractors

Whilst it is not easy to make comparisons between British firms because of their widely diverse activities, this is easier than comparing companies which cross national boundaries. All sorts of hidden or unquantifiable factors need to be considered such as culture and accounting systems that have evolved to meet different objectives [4].

In Britain annual financial accounting is prepared with the intention of providing the shareholders in a company with a true and fair view. This means that a company's accounts should allow the reader to gain a clear understanding of the position of the business in financial terms. Thus the accounts are normally prepared on an income and expenditure basis to allow revenues to be matched with related costs. On mainland Europe, accounts are almost solely undertaken for taxation purposes and are thus prepared on the basis of minimizing reported profits, often resulting in the purchase of assets, which then result in a lower tax bill.

Company ownership is also important. In Britain all large contractors are owned by shareholders who use dividend growth, i.e. earnings per share, as the yardstick of their support. Shareholders often take a short-term view of company performance. Companies that do not achieve adequate growth become ripe for a takeover. In the majority of the other European countries, Netherlands being the exception, which adopts a similar system to that of Britain, most of the companies are owned by long-term investors. In Germany the banks provide the finance and also have a large stake in the companies. In Italy many of the companies are still family owned. Tax efficient assets rather than dividend growth is used as the yardstick.

However, in attempting to make comparisons, the French firm Bouygues continues to remain the largest contractor in Europe (Table 6.8). The accounting anomalies described above and the fact that many construction firms are now widely diversified across different markets adds to the complication. Almost gone are the days when a company undertook a single line of business. For example, 10% of Bouygues' turnover is achieved through a French television station. In order to qualify on this list as a contractor, more than half of the group's turnover must come from contracting. Much of the BICC's activities come from cabling activities but its construction division, Balfour Beatty, achieved more than half of the group's turnover.

The larger French and German construction groups now dominate the European scene in terms of their size. The largest British firm, at the time was Tarmac (see Table 6.7) at eleventh position in the table. In 1989, it was fourth. Also whilst Bouygues saw its turnover double from £4500m in 1989 to £9312m in 1994, Tarmac's, over the same period of time, declined from £3520m to £2515m. This was symptomatic of the poor state of the British construction industry in the early 1990s. Amongst the top twenty European contractors, only two are now British, the other being AMEC.

Bouygues is planning to gain a foothold into the British construction market, to add to its existing overseas work in Germany, Belgium

Table 6.8   The top ten European contractors [2] [4]

Rank	Company	Country	Turnover (1994) (£m)	Turnover abroad %	Profit (1994) (£m)	No. of Employees
1	Bouygues	France	9 312	30	74	91 251
2	SGE	France	5 891	44	37	63 366
3	Holzmann	Germany	5 790	31	47	43 264
4	Hochtief	Germany	4 625	34	57	35 382
5	Eiffage	France	4 221	16	40	43 040
6	GTM-Entrepose	France	3 972	37	26	68 201
7	Bilfinger & Berger	Germany	3 379	42	57	47 071
8	Dumez-GTM	France	3 183	46	8	33 900
9	Skanska	Sweden	2 859	34	265	28 868
10	Strabag	Germany	2 691	2	18	23 385

and Spain. Ballast Needham, a Dutch contractor bought Wiltshier in 1995 and Tilbury Douglas is already 30% owned by Holzmann, the German contractor. With a few exceptions, British contractors have still to make their move across the Channel into mainland Europe.

Japanese involvement in Great Britain

The Japanese contractors dwarf those of Europe. By comparison the largest of these is almost three times the size of Bouygues, the biggest European firm. However, their turnover in Britain is tiny, by comparison with the indigenous firms (Table 6.9). The big three Japanese firms in Europe [6] all have their main headquarters located in London, but they each also have a presence in the capital cities of the main European countries. About one-third of their staff are currently Japanese. Some have forged strong links with developers but none at the moment, apparently, want to become developers in their own right, but would rather carry out aspects of the project with which they are more conversant and capable.

Table 6.9   The big three Japanese contractors in Europe; basic facts, 1991 [6]

	Shimizu	Taisei	Kajima
Established	1804	1873	1840
Order book 1989/90	£8.3bn	£7.6bn	£8.4bn
Number of employees	10 600	9 000	13 200
Set up in GB	1979	1984	1985
Number of GB employees	114	20	88
GB turnover	£70m	na	£50m
Other in employees in Europe	50	20	60
European Headquarters	London	London	London

Some British Contractors

John Laing PLC

The origins of the John Laing Group of companies can be traced to a small house in the Lakeland village of Sebergham. The house, which is still standing, was built by James Laing who was a son of a Scottish master builder and mason. His eldest son, John, expanded the business and opened an office in Carlisle in 1867. In 1920, the company became a limited company and in 1926 moved to its present headquarters at Mill Hill. The company played a major part in the post Second World War building boom, being heavily involved in house building, completing the first part of the M1 motorway and Coventry Cathedral. The company has maintained its policy of social awareness, which has been a hallmark since the time of its founder. The company employs about 12 900 (5700 staff and 7200 operatives) both at home and overseas, and has a turnover of £1.6bn. More recent projects include Sizewell 'B' Power station, the second Severn crossing and the Toyota manufacturing plant. In addition to Laing-named companies, J L Project Management Limited and Elstree Computing Limited are a part of the group.

George Wimpey PLC

In 1880 a young man of 25 called George Wimpey established a stone-working business in partnership with a mason, Walter Tomes. Their yard in Hammersmith Grove was ideally situated close to Grove Road station, which served both the Great Western and the London and South Western railways. As business expanded materials could be easily received from quarries from all over Britain for masonry of all kinds. In 1893 George Wimpey became the sole proprietor. After the Second World War the company took shape as a worldwide organization and became a household name. The company became well known as a house builder and main contractor both for building and civil engineering projects. It has also developed activities in minerals-related operations which includes the quarrying of aggregates, the production of asphalt and concrete products. The group has an annual turnover of £2bn worldwide and provides employment for more than 15 000 people in all aspects of construction across six continents. Its activities are focused in four core businesses: homes, contracting, minerals and property. Some of its notable projects have included: Hong Kong and Shanghai Bank, a part of British Nuclear Fuels at Sellafield, Highland One North Sea oil rig and the Coppergate Centre in York. In 1996, Wimpey withdrew from contracting to become the UK's largest house builder with about 8% of the market.

Bovis Construction Group

The Bovis Construction Group is a major division of The Peninsular and Oriental Steam Navigation Company. This is one of the world's most successful shipping, transport, property, and construction companies. P&O was founded in 1837 and has its origins in the days when Britain's merchant and passenger shipping fleets first spanned the world. In 1927 Bovis Construction devised a fee system of building for a new client, the fast growing Marks & Spencer. Based upon this system Bovis developed its own management contracting techniques and its reputation in this field of activity. In 1991, the house building, construction and development division had a turnover of £1.79bn of which 49% related to work in the United Kingdom and Eire. Some of its notable projects have included: Lloyds Building, Canary Wharf, Broadgate offices, Meadowhall shopping centre in Sheffield and the rebuilding of Strangeways Prison in Manchester.

References

[1] *Housing and Construction Statistics*. HMSO
[2] Building 500, Annual Survey. *Building* magazine 1995
[3] Cooper P. and Savin J. 'Ranking stability' in *Building*, 16, November 1990
[4] Coles D. and Singleton K. 'Accounting for differences' in *Building*, 16, November 1990
[5] 'The top 500 European companies', annual survey in *Building*
[6] Ridout G. 'Oriental Express' in *Building*, 24 May 1991
[7] Latham, M. *Constructing the Team*. HMSO 1992

7

Procurement

Suppose one of you wants to build a tower. Will he not first sit down and estimate the cost to see if he has enough money to complete it?

Luke 14:28

Introduction

Traditionally a client who wished to have a project constructed would invariably commission a designer (an architect for building projects or an engineer for civil engineering projects) to prepare drawings of the proposed scheme. Where the project was of a sufficient size a quantity surveyor would be employed to prepare estimates and documentation on which the contractor could prepare a price. This was the procedure used during the early part of the century. Even up to 30 years ago there was a limited choice available of methods used for procurement purposes. Since the early 1960s there have been several catalysts for changes in the way that projects are procured:

- Government intervention through committees such as the Banwell Reports.
- Pressure groups being formed to create beneficial change for their members, most notably the British Property Federation.
- International comparisons, particularly with the USA and Japan, and influence of the Single European Market in 1992.
- The apparent failure of the construction industry to satisfy the perceived needs of its customers, particularly in the way in which it organizes and executes its projects.
- Influence of educational developments and research.

Figure 7.1   *Innovative schemes are difficult to cost forecast*

- The response, particularly in times of slumps in the industry, towards greater efficiency.
- Changes in the ways in which technologies are used and attitudes amongst the professions.
- The clients' desire for single point responsibility.

However, there is no panacea and procedures continue to evolve in order to meet new situations. Hybrid developments are being sought to amalgamate the best from the competing alternatives. The different methods have all been used at some time in the industry, some more than others due largely to user familiarity and ease of application and recognition. New methods will continue to be devised to meet new requirements and demands from clients, contractors and the professions.

Appetite for change

Table 7.1 highlights the interest in the development of alternative procurement and contractual systems which are used in the building industry [1]. It has been derived from the number of published papers concerned with these alternative systems. It excludes the large

Table 7.1   Appetite for change in procurement methods [1]

Year	No. of published papers	Year	No. of published papers
1968	3	1978	34
1969	5	1979	30
1970	15	1980	23
1971	8	1981	39
1972	14	1982	31
1973	0	1983	34
1974	18	1984	33
1975	30	1985	33
1976	25	1986	33
1977	46	1987	33

numbers of books and chapters in books, seminars and conferences which have taken place over the years on this subject. More recently an International Journal of Construction Procurement, with an editorial board represented by individuals from several different countries is now being published four times per year. The international research organization, CIB, has a Working Commission (No. 69) that is solely devoted to procurement.

Constructing the team (the Latham Report)

There have been numerous reports on the construction industry, many of which have been Government sponsored, that have been published over the past fifty years. Their overall themes have been aimed at improving the way that the industry is organized and the way that construction work should be managed. Value for money has been a major theme and procurement an important issue. These reports have ranged from the Simon Report of 1944 (The Placing and Management of Building Contracts) to the more recent (1996) report on design and build published by the Centre for Strategic Studies in Construction at the University of Reading (Designing and Building a World Class Industry). This university also published the report, Building Towards 2001 (1991). The well-known government sponsored report on procurement, known as the Banwell Report (The Placing and Management of Contracts for Building and Civil Engineering Works) was published in 1967. In 1994, a committee under the chairmanship of Sir Michael Latham MP published the far-reaching report, Constructing the Team, which is commonly referred to as the Latham Report.

The Latham Report provides a review of the construction industry. It was jointly commissioned by the government and the industry. It also included contributions from various client bodies on their expectations when undertaking major capital works programmes.   The Joint Review of Procurement and Contractual Arrangements in the United Kingdom Construction Industry was announced in the House of Commons in 1993. An interim report, titled, Trust and Money was published later that year. The funding parties to the Review included the Department of the Environment (DoE), the Construction Industry Council (CIC), the Construction Industry Employers Council (CIEC), the National Specialist Contractors Council (NSCC) and the Specialist Engineering Contractors Group (SECG). Clients have continued to be closely involved in the Review and were represented by the British Property Federation (BPF) and the Chartered Institute of Purchasing and Supply (CIPS).

The Interim Report sought to:

- describe the background to the Review and its parameters
- define the concerns of the differing parties to the construction process, some of which were mutually exclusive
- pose questions about how performance could be improved and genuine grievances or problems assessed
- reiterate and expand upon what has long been accepted as good practice in the industry, but is often honoured more in the breach than in the observance

Whilst the response from the different sectors of commerce and industry has been received with varying degrees of enthusiasm and cynicism the resultant changes in practice has been more attributed to pragmatic courses of action and financial pressures rather than necessarily the advice offered by the originators or committees who prepared the reports. Some of the recommendations of the different reports have been implemented. Other problems and difficulties continue to persist, even though the structure of the industry and the nature of its clients have changed considerably. The construction industry comprises many different parties, organizations and professions. These represent many with vested interests and traditions that in some cases represent power and authority. Such bodies are clearly loathe to relinquish these positions freely.

Clients remain at the core of the process and their needs must be met by the industry. Implementation, after all begins with them. Clients are also dispersed and vary greatly. Government, in the past used act as a monolithic client. The privatization of many government departments and activities have changed this perception, resulting in the fragmentation of this important client base. Some government agencies are now indistinguishable from the private sector. Even the existing government departments now operate different procurement strategies and practices. This became more pronounced after the demise of the Property Services Agency (PSA).

The Latham Report made to reference to a wide range of different problems. The following represent some of the major issues.

Fair construction contracts

The Construction Sponsorship Directorate of the Department of the Environment published a consultation paper under the above title in May 1995. The review recognized that present arrangements used in the industry mitigated against co-operation and teamwork and therefore against the clients of the industry. These also contribute towards helping to perpetuate the generally poor image of the construction industry.

This consultation paper identifies four areas which it describes as essential terms in any fair construction contract:

Dispute resolution: emphasizing the need to adopt ADR (Alternative Dispute Resolution) procedures (see Chapter 5)

Right to set off: This principle is widely applied beyond the remit of the construction industry. The principle is as follows: where A claims money from B, B is entitled to deduct any money that A owes before paying the balance. The Latham Report recommends that the right to set-off should be restrained as follows:

- a requirement to give advance notification (with reasons) of the intention to apply set-off
- set-off subject to adjudication
- set-off only to be allowable in respect of work covered by the contract

Prompt payment: with added interest in the case of default

Protection against insolvency: the use of trust funds (see below)

This report summarizes what it considers are the fundamental principles of a modern contract. The points made are as follows:

- dealing fairly with each other and an atmosphere of mutual co-operation
- firm duties of teamwork, with shared financial motivation to pursue those objectives

- an interrelated package of documents, clearly identifying roles and duties
- comprehensible language with guidance notes
- separation of the roles of contract administrator, project or lead manager and adjudicator. The project manager should be clearly defined as the client's representative
- allocating risks to the party best able to control them
- avoiding the need, wherever possible, of changes to pretender information
- assessing interim payments through milestones or activity schedules rather than through monthly measurement
- clearly setting out the periods for interim payments and automatically adding interest where this is not complied with
- the provision of secure trust funds
- provision of speedy dispute resolution
- provision of incentives for exceptional performance
- making provision for advance payment to contractors and subcontractors for prefabricated off-site materials and components

Trust funds

It is fundamental to trust within the construction industry, that those involved should be paid the correct amounts at the right time for the work that they have carried out. It may be argued that a problem does not exist and that:

- clients only award work to firms with integrity
- contractors are at liberty to decline work from dubious clients
- subcontractors can adopt similar business practices
- bonds and indemnities are already available
- bad debts are not singularly a problem in the construction industry

However diligently clients, contractors and subcontractors verify each other, the realities of the construction industry and its markets continue to exist. In circumstances, such as a recession, contractors and subcontractors are almost prepared to undertake work for any client. This is frequently done at a minimal profit margin. Bad debt insurance is available but adds extra costs at times when firms are seeking to reduce overheads. In times of prosperity, clients are prepared to undertake work with almost any firm who is available in order to get an important project constructed.

The contractor's goods and services become part of the land ownership once incorporated within the project. Any 'retention of title' clause that might be incorporated by suppliers or contractors in their trading agreements does not protect them once the materials are incorporated within the works. The building contractor is also likely to be sufficiently far down the queue, where an employer is unable to make payment within the terms of the contract. In some countries around the world, legislation has been provided to deal with the potential injustice that might be suffered. The most comprehensive is the 'Ontario Construction Lien Act 1993'.

An effective way of dealing with this problem is to set up a trust fund for interim payments and retention monies. A client, for example, could be requested to pay into such a fund at the start of the payment period, e.g. at the beginning of the month. The correct payment, duly authorized, would then be paid to the contractor at the appropriate time. Where a form of stage payments was used, then the amount of the particular programme stage would be deposited in the trust fund at the commencement of the work in this stage. Where a bill of quantities was used then the client's quantity surveyor would request an appropriate amount of money to be placed in the trust fund to cover the next monthly certificate. The amounts authorized should correspond with the contractor's approved contract programme. The main contractor and the subcontractors would be informed of the amounts deposited. If any party considers that the sums are inadequate then they should have the option of applying to the adjudicator for its increase. There may be some argument of making payments to the subcontractors directly from this fund rather than through the main contractor's account.

Whether trust funds should be provided for all contracts is a matter of some debate. Some in the industry suggest that trust funds should only be available for projects with long contract periods or for contracts of a certain monetary size. The figure suggested for the latter is £0.25m. On very small projects the agreement is often reached about making no payment until all of the work is fully completed.

Any monetary interest accrued in the trust fund belongs to the client. Where the fund necessitates bank charges, then these costs need to be determined at the time of tender.

Such funds are not really required for public works projects, since it is not likely that such bodies will become insolvent. However, such funds will be a source of reassurance for subcontractors, if the main contractor becomes insolvent during the course of the work. If trust funds are to be used in the construction industry then they should be used in the public as well as the private sector.

Project partnering

During the 1980s the United Kingdom construction industry suffered one its most severe recessions this century. Many of the largest and best contractors and design firms made losses in successive years and many of the smaller and not the worst firms ceased to trade.

There are many in the construction industry who now recognize the need to move away from the confrontational relationships which cause the majority of disputes, problems, delays and ultimately expense.

The definition of partnering proposed by the National Economic Development Council, is a long-term commitment between two or more organizations for the purpose of achieving specific business objectives by maximizing the effectiveness of each participant's resources.

The Banwell Report (1967) recognized that there was scope for the awarding of contracts without the use of competitive tendering under certain circumstances, for example, where a contractor had established a good working relationship with a client over a period of time, completing projects within the time allowed, at the quality expected and for a reasonable cost. These circumstances may have arisen through serial contracting and particularly in those situations of continuation contracts where projects had been awarded on a phase basis. In the private sector, project partnering has been carried out successfully by some clients and contracting firms for many years. One of the claims of Bovis, the construction firm, is that it never seeks work through competitive tendering. Many large industrial corporations, also regularly use and favour those firms who they believe offer the best all-round solution to their building needs.

However, there has always been some reluctance on the part of public bodies to adopt such procedures, since they may lack elements of accountability, even though it could be shown that a good deal had been obtained for the public sector body concerned. Public sector bodies also need to follow EU procedures where appropriate. In 1995 a Government white paper on purchasing gave the official seal of approval to partnering in the public sector. This provides an environment that is similar to the private sector approach with a number of construction firms developing long-term relationships with public clients. Effectively partnering is possible where:

- it does not create an uncompetitive environment
- it does not create monopoly condition
- the partnering arrangement is tested competitively

- it is established on clearly defined needs and objectives over a specified period of time
- the construction firm does not become over-dependent on the partnering arrangement

Partnering is well established in the USA and has been used in the UK successfully by a number of different major companies such as Marks & Spencer, Rover Group, British Airports Authority and Norwich Union. The partnering arrangement may last for a specific length of time or for an indefinite period. The parties agree to work together in a relationship of mutual trust, to achieve specific objectives by maximizing the effectiveness of each participant's resources and expertise. It is the most effective on large construction projects or in circumstances where repetition occurs on projects that are constructed in different locations. The McDonald's chain of restaurants provides a good example of the latter. This arrangement, coupled with innovative construction techniques, has helped towards reducing the costs and time of construction by 60% since the start of the 1990s. Its on-site activities have been reduced from 155 days to 15 days for a typical restaurant building.

The concept of Partnering extends beyond the client, contractor and consultants and includes the different subcontractors, specialist and suppliers involved with the project. The companies work together as a joint company. It is therefore suitable for design and build arrangements.

Cost reductions

One of Latham's important recommendations was to achieve a 30% reduction in building costs by the year 2000. Such cost reductions should not generally reduce quality, unless there are issues of over-specification. Also whilst costs decrease, quality and value should at least be maintained, and in the light of some of the above comments they also need to be improved. At least one major client of the construction industry, British Airports Authority, has already intimated that the cost-value reductions proposed will be insufficient to meet their general aims and objectives. However, most industry commentators feel that 30% is too optimistic and is unlikely to be achieved on a large scale, even if the industry was radically reorganized.

It is especially important that any cost reductions in buildings does not refocus the construction industry backwards fifty years towards the emphasis on initial costs alone. The importance of ensuring that life cycle costs are given their rightful importance in the overall building process must be maintained.

The implications of reducing the costs of buildings, is one of adding value. It is the principle of doing more with less, an aim that is now prevalent throughout many areas of society. Before examining possible areas where cost reductions might be possible, the following points should be considered in context.

General issues

Recession

The construction industry in Great Britain has been in a severe recession, some will argue the worst recession for the industry this century. As any economist knows, prices charged for a commodity relate basically to the supply and demand of that commodity. The construction industry's products are no different in this respect. Any anticipated real cost reduction in building costs, over the next five years, will therefore need to take into account the fact that present costs are being artificially depressed due to the recession. Therefore it needs to be asked whether such cost reductions will be real or only relative?

Low wage industry

The industry is a low wage industry when compared with other industries. Any comparison of the wages and salaries paid at all levels in construction with industries such as energy or air transport, indicates this fact. The industry has also been described by some as a 'handicraft' industry because of its lack of use of modern technologies. Since it is not a high wage industry it frequently does not encourage the best individuals to join it, at any level. If the industry wants more innovation and efficiency then it needs to attract better talent at all levels. This inevitably means increasing salaries and wages, and hence costs, at least initially.

Government involvement

The argument is often used to suggest that the industry is inefficient and clings to out-dated practices. The industry also fails to invest properly due to its focus on short-term gains. The reason for these are obvious. The booms and slumps in the industry help to encourage and prolong such inefficiencies. Government could alter its own methods and timing of construction procurement. It could choose not to build during times of prosperity, but only build when private industry is in recession. It would then pay less for its projects, provide more stable prices and be able to offer a more even workload to the industry. Instead of using the industry as a regulator for ill, it could be used for gain.

Training and education

Contractors include in their tenders substantial sums of money to rectify poor quality work on site. Training is task driven and developing the skills to carry out tasks. Education is much broader and attempts at changing the attitudes and cultures of individuals. The current emphasis at building craft level is training specific, with little attention being paid towards aspects of education. This does not engender pride in the work being carried out. This coupled with the output-based incentive schemes, results in a poor product, expensive remedial work and dissatisfied clients. The industry has a poor reputation; better education and training could assist in improving this image.

Already achieved cost efficiencies

It must also be accepted that substantial cost efficiencies have already been introduced and maintained in many different ways during the latter half of the twentieth century through, for example, cost planning, mass production, bulk purchase, prefabrication, buildability, subcontracting, off-site manufacture, etc. This, of course, in no way mitigates against a view that further cost reductions are not therefore possible, even in these areas of activity. In fact just the opposite is true. Since reductions have already been achieved in the past, this encourages the possibility for the future. Table 7.2 lists some of the progress already achieved in cost-value efficiencies.

Any comparison between buildings constructed at the beginning and end of this century also provide evidence that considerable

Table 7.2   Examples of progress achieved in cost-value efficiencies

Use of artificial rather than natural materials
Prefabrication off-site of building components
Standardization of designs and components
Simplicity in design and detailing
Selection of appropriate procurement methods
Buildability
Efficiency through engineering technology
Construction management on site
Integrated design and construction

progress in the reduction of building costs has been achieved in a number of different areas. (See *Cost Studies of Buildings*, second edition 1994 by Allan Ashworth.) In essence, it is unlikely that we could now afford to build housing, schools and hospitals in the manner constructed at the turn of the century. Many modern designs incorporate improved construction and technology at lower costs, whilst still achieving the same broad client objectives in terms of function and appearance. The examination of such buildings help to illustrate the principle of maintaining or increasing value, whilst at the same time reducing cost.

Cost-value reduction

The following are some of the issues associated with the construction of buildings that can effect possible cost-value reductions.

- Complete design at tender stage
- Teamwork
- Construction as a manufacturing process.
- Increased standardization and prefabrication
- reduce changes to the design
- optimize specifications
- improve design cost effectiveness
- apportion risk efficiently
- improve productivity
- reduce waste
- use cost-efficient procurement arrangements
- improve the use of high technology for both design and construction

Private Finance Initiative (PFI)

The purpose of this initiative is to encourage partnerships between the public and private sectors in the provision of public services. The scheme is outlined in 'Private Finance and Public Money' (Department of the Environment, 1993). In 1992, the Chancellor of the Exchequer announced a new initiative to find ways of mobilizing the private sector to meet needs that had traditionally been met by the public sector. Achieving an increase in private sector investment will mean that more projects will be able to be undertaken. This takes into account the Government's objective that public spending should decline in the medium term. The broad aims of such a partnership will be to:

- achieve objectives and deliver outputs effectively
- use public money to best effect
- respond positively to private sector ideas

In exploring the possibilities for private finance, including proposals from the private sector, the options being considered include:

- can the project be financially free standing?
- is it suitable for a joint venture?
- is there potential for leasing agreements?
- is there potential for Government to buy a service from the private sector?
- can two or more of these elements be brought together in combination in any particular instance to form innovative solutions?

Concessionary contracting falls neatly into such an arrangement, whereby the private sector is encouraged to construct public projects, such as roads and then charge a levy on this provision, for a fixed period of time specified in the contract. The contractor throughout the entire period is responsible for the maintenance of the works. Upon eventual handover to the public sector the contract will also specify the required condition of the asset.

General matters

Consultants vs. contractors

The arguments for engaging either a consultant or a contractor as the client's main advisor or representative are to some extent linked with tradition, fashion, loyalty, dissatisfaction and the belief that an alternative approach to procurement will solve most of the problems which can arise. The emphasis on a single point responsibility for the client does not automatically mean design and build, but a re-evaluation of existing arrangements. The following factors need to be considered:

- single point responsibility
- integration of design and construction
- premier client interest
- impartial advice
- needs for inspection, payments, warranties
- overall costs of design and construction

Competition vs. negotiation

There are a variety of different ways in which businesses, whether designers or contractors, can secure work or commissions. These include: invitation, recommendation, reputation and speculation. However, irrespective of the final contractual arrangements which are chosen by the client, both a designer and a contractor have to be appointed. The choices which are available for this purpose are either competition or negotiation. All of the evidence which is available suggests that the client, under normal circumstances of contract procurement is able to strike a better bargain if some form of competition on price, quality or time exists. There are, however, circumstances under which a negotiated approach with a single firm may prove overall to be more beneficial to the client. Some of these include:

- business relationship
- early start on site
- continuation contract
- state of the construction market
- contractor specialization
- financial arrangements
- geographical area

Whilst the above list is not exhaustive it should not necessarily be assumed that negotiation will always be preferable in the above circumstances. Each individual project needs to be examined on its own merits and a decision taken based upon the particular circumstances concerned and the advantages to the client.

Where competition is envisaged then a choice between selective or open competition has to be made. Selective competition is the traditional and most popular way of appointing a contractor. In essence, a number of firms of known reputation are selected by the client's advisors to submit a price. The firm which submits the lowest price is usually awarded the contract. With open competition the details of the proposed project are advertised, and any firm which feels able to complete the project within the stipulated conditions is able to submit a tender. The use of open tendering may remove from the client the moral obligation of accepting the lowest price since firms are not normally vetted before tenders are received, and factors other than price must also be considered. With selective competition the number of tenders is restricted to about six firms. With open tendering there is, in theory, no limit to the number of tenders and exceptionally, there are records of almost 100 firms submitting a tender for the same project! It should be noted that the preparation of tenders is both expensive and time consuming and that these costs must be borne by the industry. There is no such thing as a 'free estimate'.

Measurement vs. reimbursement

There are essentially only two ways of calculating the costs of construction work. One is on the basis of paying for the work against some predefined criteria or rules of measurement. The alternative is to reimburse the contractor the actual costs involved in construction and to use a system of reimbursement. Measurement contracts distribute both more risk and incentive to the contractor to complete the works efficiently. Reimbursement contracts result in the contractor receiving only what is spent plus an agreed amount to cover profits. The rates or prices quoted by a contractor on a measurement contract must allow for everything which is contained in the contract. The points to be noted in this choice are as follows:

- Cost forecast: other than in the broadest of terms this is not possible with reimbursement contracts.
- Contract sum: reimbursement contracts cannot provide a contract sum.
- Efficiency: cost reimbursement contracts often encourage the contractor to be inefficient.
- Price risk: measurement contracts allow for this. Employers may therefore pay for non-events.
- Cost control: there is little control of construction costs on a reimbursement contract.
- Administration: reimbursement contracts require a large amount of clerical work and record keeping.

Traditional vs. alternatives

Until recent times the majority of the major building projects were constructed using single-stage selective tendering. The British construction industry had evolved this system within the parameters of accepted principles and procedures. However, in more recent years, due to the possibility of new approaches towards procurement and a better knowledge of practices around the world, the traditional method of procurement has been under reappraisal for the following reasons:

- appropriateness of the service provided
- length of time from inception through to completion
- projects over-running their contract periods

- final costs being higher than expected
- problems of quality control
- mismatch between design and construction
- line of legal responsibility
- limitation on the procurement advice available

Alternative procurement procedures have therefore been devised in an attempt to address the issues identified above. None of the newer contractual procedures singularly address all of these criticisms. Procurement practice is really a case of trading off different methods against client objectives in an attempt to achieve a best possible solution.

Procurement strategy

The selection of appropriate contractual arrangements for any but the simplest type of project is difficult owing to the diverse range of options and professional advice which is available. Much of this advice is also in conflict and lacking a sound research basis for evaluation. For example, a design and build contractor is unlikely to recommend the use of an independent designer. Such organizations believe that the full integration of design with construction is likely to achieve the best long-term solution both for the project and the client. The professions which provide only a design service generally take the opposite viewpoint. Individual experiences, prejudices, vested interests, familiarity, the need and desire for improvements are factors which have helped to reshape procurement in the construction industry. The proliferation of differing procurement arrangements have resulted in an increasing demand for systematic methods of selecting the most appropriate arrangement for a particular project [2]. The reality of the situation is that a particular project with defined objectives will result in the selection of the appropriate procurement options. Table 7.3 identifies factors which should be considered when choosing the procurement path.

Methods of price determination

Building and civil engineering contractors are paid for the work which they carry out on the basis of either measurement or reimbursement of costs.

Table 7.3   Procurement selection

Size: small projects not suited to complex arrangements

Design: aesthetics, function, maintenance, buildability, contractor integration, bespoke design, design before build and prototypes.

Cost: price competition/negotiation, fixed price arrangements, price certainty, price forecasting, contract sum, bulk-purchase agreements, payments and cash flows, life cycle costs, cost penalties, variations and final cost.

Time: inception to handover, start and completion dates, early start on site, contract period, optimum time, phased completions, fast tracks, delays and extension of time.

Quality: quality control, defined standards, independent inspection, design and detailing, single and multiple contractors, co-ordination, buildability, constructor reputation, long-term reliability and maintenance.

Accountability: contractor selection, *ad hoc* arrangements, contractual procedures, auditing, simplicity, value for money.

Organization: complexity of arrangements, standard procedures, responsibility, subcontracting and lines of management.

Risk: evaluation, sharing, transfer and control.

Market: workloads, effects of procurement advice.

Finance: collateral, payment systems, remedies for default and funding charges.

Measurement

Using this method the work is measured on the basis of the actual quantities which are incorporated into the finished project. The contractor is paid for this work on the basis of quantity multiplied by rate. (The contractor is also paid hire charges for plant, temporary materials, etc.) The work may be pre-measured for tendering purposes in which case it is known as a lump sum contract. Alternatively it may be measured after being executed for payment purposes, in which case it is known as a remeasurement contract. On most projects, however, some form of remeasurement will occur to take account of variations. Building contracts are more often lump sum contracts, whereas civil engineering contracts are typically of the remeasurement type. The following are the main types of measurement contracts which are used:

Drawings and specification.   This is the simplest type of measurement contract and is really only suitable for small or straightforward projects. Each contractor has to measure the quantities from the drawings, interpret the specification and prepare a tender price.

Performance specification.   With this method the contractor is required to prepare a price on the basis of the client's brief and user requirements alone. The contractor is left to decide upon the type of materials and construction which meet with the performance criteria.

Schedules.   These form two categories. A schedule of rates, which is similar to a bill of quantities but with an absence of any quantities. A schedule of prices which is an already-priced document, where the contractor's tender is simply a percentage adjustment to these prices. Neither of these methods are able to calculate a contract sum.

Bills of quantities.   The contractors' prepare their tenders on detailed quantitative and qualitative data. The information is prepared in accordance with some agreed rules of measurement, e.g. Standard Method of Measurement of Building Works (SMM7), Civil Engineering Standard Method of Measurement (CESMM3), etc. The contractors tendering for the project are all able to use the same detailed information for pricing purposes. In circumstances where the work is unable to be properly defined, bills of approximate quantities may be prepared.

Cost reimbursement

On these types of contract the contractor is able to recoup the actual costs of materials which have been used and the time spent on the project by the operatives. An amount is agreed to cover the contractor's profit. Dayworks are valued on a similar sort of basis. Because these types of contractual arrangement are unable to provide a clear indication of expected costs and offer little incentive for the contractor to control costs, they should be used only in special circumstances:

- emergency works projects
- where the character and scope of the works cannot be easily determined
- where new technology is being used

- where a special relationship exists between the client and the contractor
- where it would be unfair to ask a contractor to price the works, for example, in cases of exceptional risk

Cost reimbursement contracts can take several different forms. The cost plus percentage arrangement, where the contractor's profits are in a direct relationship with costs, i.e. the higher the costs the higher the profits. The cost plus fixed fee, where the fee is fixed beforehand and does not alter with a rise or fall in costs. A third method is the cost plus variable fee which supposedly provides some incentive on the part of the contractor to control costs. The difficulty with these latter two methods is related to the general inability to be able to accurately forecast costs in advance on which the fixed and variable fees are to be calculated.

Price forecasting

The estimate or forecast of cost of construction is done at different stages of the project. A budget sum for the client on behalf of the design team, or design-and-build contractor, is prepared at the inception. If this is acceptable then more detailed methods are used to provide a framework for cost planning. Eventually the contractor estimates the costs in order to calculate the tender sum. If successful then this sum is incorporated in the contract documents as the contract sum.

Design cost estimating uses a variety of techniques (Table 7.4) which have become known as single-price methods, even though in some cases they use a limited number of cost descriptors. These methods can be no better than the information on which they are based. Table 7.5 provides a list of typical building costs based upon the superficial floor area method of calculation. The actual rates vary due to shape, size and standards of construction, finishings and engineering services that are included. The market condition at the time of tender is also an important factor.

In most cases during the early part of the design stage, the drawings and specifications are uncertain and imprecise. However, a forecast of costs is still required. As the design progresses the forecast can be refined, but on average it will vary by at least 10% from the eventual contract sum, which in itself is still an estimate of cost. It is usual to offer a range of possible estimates or confidence limits rather than a single sum. There is a large amount of data available which measures pre-tender estimates against final costs. The relationship is often poor but this belies factors which are outside the control of the cost advi-

Table 7.4   Methods of pre-tender estimating [3]

Method	Description
Conference	Based on a consensus view
Financial methods	Cost limits determined by the client
Unit	Used on projects having standard units of accommodation
Superficial	Based upon floor area
Superficial-perimeter	Based upon a combination of floor area and the building's perimeter
Cube	Based upon the project's volume
Storey enclosure	Based upon a combination of weighted floor, wall and roof areas
Approximate quantities	Major items measured
Elemental estimating	Used in conjunction with cost planning
Resource analysis	Used mainly by contractors
Cost models	Computerized systems using mathematical formulae

sor. The future is always difficult to forecast and errors and inconsistencies in pricing occur. Table 7.6 is a list of different projects, comparing their initial cost estimates with their predicted final costs. Whilst the forecasting of construction costs is poor, this is a worldwide problem rather than one restricted to the British construction

Table 7.5   Typical building costs per square metre of floor area [4]

	$£/m^2$
Car parks	121–256
Offices	436–1039
Shops	266–782
Schools	405–844
Factories	191–494
Banks	644–1156
Hospitals	573–1120
Churches	498–974
Universities	421–997
Houses	281–525

Table 7.6   Some examples of estimating accuracy

	Estimate (£m)	Actual (£m)
Sydney Opera House	2.5	87
Thames Barrier	23	461
Barbican Arts Centre	17	80
Natwest Tower	15	115
Humber Bridge	19	120
Concorde	250	1200
British Library	164	450
BNFL-Thorp	300	2800
Palace of Westminster (1834)	0.72	1.5
Eurotunnel	4.7bn	10bn+
British Library	164m	496m
House for Mr Bill Gates (Microsoft)	10bn$	30bn$
Apollo space programme	12bn$	21bn$
Trans Alaskan pipeline	900m$	8.5bn$

industry alone. The table also indicates that other industries fare no better. The list could be almost endless, and have included other projects which had large cost over-runs, such as: RB211 programme, Advanced Passenger Train (which was finally abandoned), North Sea oil platforms, Heysham II, Polaris and the computerization of PAYE. All of these projects cost much more than was originally expected. An estimate, by definition, will always be subject to some form of error. Even the Duchess of Windsor's jewels which were estimated to be worth £5m were sold at auction for £35m! The Walt Disney film 'Snow White' was budgetted to cost 250 000$, but its final cost was in excess of 1.4m$.

Table 7.7   Estimate classification and accuracy [5]

Estimate	Purpose	Accuracy
Order of magnitude	Feasibility studies	+/− 25–40%
Factor estimate	Early stage assessment	+/− 15–25%
Office estimate	Preliminary budget	+/− 10–15%
Definitive estimate	Final budget	+/− 5–10%
Final estimate	Prior to tender	+/− 5%

Some of the above differences in costs can be accounted for by changes in design and specification, changes in client's requirements, the introduction of new technologies and inflationary factors. Also many of these projects represented innovatory solutions, in terms of design and technology, which required considerable modifications during construction. The usual levels of cost estimating accuracy that are achieved in practice are shown in Table 7.7. These can be applied to the full range of different types of construction projects.

Contractors' estimating is based upon measuring a large number of work items and analysing their unit costs, using previously recorded site performance data. The measured items should aim to consider only the cost-important items. In theory the site performance data or labour outputs and material and plant constants are retrieved from work done on site. Research has shown that this theoretical concept is flawed in practice due to the poor and inappropriate recording systems used by contractors and the lack of confidence that estimators have in individual site feedback. The time taken to undertake the different construction operations is also highly variable. The difficulty of capturing this data in a meaningful form that can then hopefully be reused is a complex task beyond the profitable occupation of most contractors. A comparison of similar items priced by different contractors reveals differences or discrepancies as high as 200%. Even published data on guide prices can vary by as much as 50% [6].

Tender prices and building costs

Several different firms and organizations within the construction industry attempt to measure changes in tender prices and building costs and to analyse the effects of regional variation on these indices. The different indices are used to measure trends in future costs and prices and have been used retrospectively by contractors to recoup increased costs with the Formula Methods of Price Adjustment used on both building and civil engineering projects.

Building cost indices

This type of index measures the contractor's costs. It is thus a combination of actual wage rates, material costs, plant and overhead charges, which have been weighted on a 'basket of goods' basis to reflect typical ratios. The representation depends upon the types and amounts of items included in the basket. The task is difficult owing to the variety of different materials and methods of construction which are used. In order to make the data more usable and informative,

Table 7.8   Typical material price indices; 1990 = 100 [7]

	1980	1985	1990	1991	1992	1993	1994	1995E
Sand & gravel	44	75	100	97	92	90	96	101
Cement (opc)	66	83	100	106	110	112	117	125
Bricks (commons)	47	78	100	99	97	102	116	122
Hardwood	49	69	100	93	97	123	140	150
Softwood	59	79	100	90	86	99	104	114
Structural steel	58	75	100	102	101	103	108	115
Copper	46	62	100	94	100	110	134	140
Plastic pipes	68	74	100	109	113	114	122	130
Sanitaryware	71	76	100	108	110	114	119	125
Insulation	67	71	100	111	118	128	138	142

different indices are prepared for the different types of building materials. Table 7.8 lists some of the more common materials together with an index of their costs from 1980 to 1995. During the 1980s, for example, cement increased in cost by 52%, bricks by 112% and softwood by 69%. During the recession of the early 1990s, these material costs continued to increase. By 1995 all material costs were higher than their 1990 prices, although in some cases this was only marginal, as in the case of sand and gravel and ready mixed concrete. These cost differences are to a large extent accounted for by supply and demand factors, raw materials prices, and because of different methods used in their manufacture. The building cost indices (Table 7.9) indicate a steady increase in material prices throughout the 1980s, materials on average being 80% more expensive at the end of the decade. The building cost indices, whilst prepared for different sorts of buildings using different combinations of materials, show a remarkable similarity in their overall change of relative price.

Tender price indices

The tender price indices are based upon what a client is prepared to pay for the project. In addition to taking into account building costs the indices also make an allowance for the contractor's profit and market conditions. Separate indices are recorded to distinguish between prices which include increases in contractors' costs and those which deal with this aspect separately. Assuming that both of these indices have a common base date, then the one which allows for fluctuations in costs to be measured separately should always display lower values. Table 7.9 shows the movement of tender prices and

Table 7.9   Building costs and tender prices [8]

		Tender prices		Building costs		Retail prices	
		Indices	Inflation	Indices	Inflation	Index	Inflation
1991	1						
	2	262	−16.6	350	6.4	335	6.0
	3	261	−16.3	360	4.0	337	5.0
	4	254	−12.4	360	3.7	340	4.4
1992	1	250	−8.1	361	3.1	342	4.3
	2	248	−5.3	362	3.4	349	4.2
	3	241	−7.7	367	1.9	349	3.6
	4	233	−8.3	368	2.2	350	2.9
1993	1	227	−9.2	370	2.5	348	1.8
	2	242	−2.4	371	2.5	353	1.1
	3	233	−3.3	373	1.6	354	1.4
	4	239	2.6	374	1.6	356	1.7
1994	1	239	5.3	375	1.6	356	2.3
	2	247	2.1	379	2.2	362	2.5
	3	266	14.2	385	3.2	363	2.5
	4	256	7.1	388	3.7	365	2.5
1995	1	258	7.9	392	4.3	368	3.4
	2	265	7.3	397	4.7	375	3.6
	3	266	0.0	407	5.7	376	3.6
	4	270	5.5	407	4.9	376	3.0
1996E	1	268	3.9	408	4.1	378	2.7
	2	268	1.2	409	3.0	384	2.4
	3	272	1.9	419	2.9	385	2.4
	4	275	1.7	421	3.4	386	2.7
1997E	1	278	4.0	423	3.7	388	2.6
	2	282	5.4	425	3.9	394	2.6
	3	287	5.9	438	4.5	395	2.6
	4	292	6.4	441	4.8	397	2.8
1998E	1	297	6.7	443	4.7	399	2.8

E = estimated

construction costs against general inflation. Whilst the construction cost indices indicate a gradual increase in line with the index of retail prices, the tender price index is much more erratic and at a much lower level. During the 1980s, tender prices increased by 48%, despite a sudden downturn at the start of the recession at the end of the decade. For example, in 1980 a project tender worth £10m would have been priced at £16.44m by 1989, but would have reduced to £14.67m by 1990. Had tender prices continued to rise in line with inflation, then the project in 1990 would have cost £17.5m.

Table 7.9 indicates that tender prices continued to fall until the third quarter of 1993. There is little correlation between building costs and tender prices even though the causal factors between them have the same origins.

Contractor selection

Contractors are selected to carry out construction work on the basis of either competition or negotiation or, in the case of two-stage tendering, a combination of both. These methods are used in conjunction with a pricing system.

Procurement options

There are a wide variety of procurement options available in the construction industry which are aimed at addressing criticisms of poor quality, long construction periods and high costs. A summary of these can be found in standard textbooks [9]. The methods vary from traditional single-stage selective tendering, where a client uses a designer to prepare drawings and documentation on which contractors are invited to submit competitive prices, to schemes where a single construction firm will provide the truly all-in service, the turnkey project. This last method might even include land acquisition and the long-term maintenance of the project. Procurement methods have been devised to get the contractor on site as quickly as possible, such as two-stage tendering and fast tracking, with of course the anticipation that the contractor will also complete the works sooner than by using the traditional approach. Other methods have recognized the contractors improved management skills and their influence on the whole construction process. Some methods have evolved to recognize changes which have occurred in the industry, such as the proliferation of subcontracting, the increase in litigation and the need for one organization to accept total responsibility for the whole process. Another method of procurement attempts to recognize that if a project is fully designed prior to tendering, then the time spent on site should be able to be shortened with a consequent reduction in construction costs. The difficulty with this approach is to combat the client's desire for work to start on-site as soon as possible. Table 7.10 identifies a number of client criteria and procurement methods which might be used to achieve such objectives.

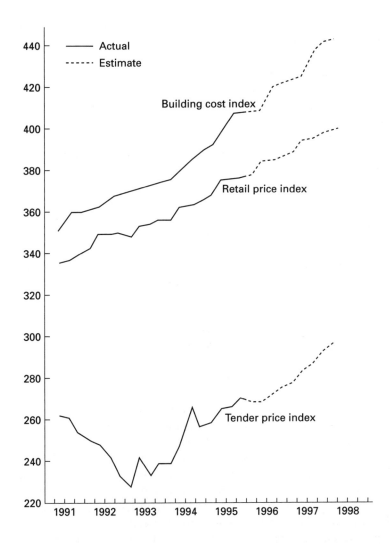

Figure 7.2   *Comparison of building costs, tender prices and retail prices*

Code of procedure for single-stage selective tendering

The National Joint Consultative Council (NJCC) has developed codes of procedures for tendering purposes which represent good practice on the awarding of construction contracts. They represent the principles of good procurement practice, many of which can be applied with the different methods which are currently available. The following are some of the more important points from the code which deals with single-stage selective tendering.

- Use a standard form with which the various parties in the construction industry are familiar.
- Restrict the numbers of tenderers. The code suggests no more than six. The costs of unsuccessful tenders will be borne by the construction industry clients.
- When preparing a short list, the following should be considered:
  - the firm's financial standing and record
  - recent experience of building over similar contract periods
  - the general experience and reputation of the firm for the type of project envisaged
  - adequacy of the contractor's management
  - adequacy of capacity.
- Each firm on the list should be sent a preliminary enquiry to determine if it is willing to tender. The enquiry should contain the following information:
  - job title
  - names of employer and consultants
  - site location together with a description of the works
  - principal nominated subcontractors
  - approximate cost range
  - form of contract
  - dates for possession and completion
  - anticipated date for despatch of tender documents.
- If it is necessary for a contractor to withdraw once he has decided to tender he should notify the client's advisor as soon as possible.
- Contractors who have shown a willingness to tender but are not chosen for the short list must be informed immediately.
- The tender period should not be less than 4 weeks but this will depend upon the size and complexity of the job.
- Where a tenderer seeks some clarification of the tender documents then all other tenderers should be informed of the decision.
- If a tenderer submits a qualified tender then this should be withdrawn or the tender rejected.
- Under English law a tender may be withdrawn at any time before its acceptance. Under Scottish law, it cannot be withdrawn unless the words 'unless previously withdrawn' are inserted in the tender after the stated period of time is to remain open for acceptance.
- All tenderers should be informed as soon as possible of the result.
- All tenderers should be provided with a list of tender prices once the contract has been signed.

Table 7.10 Identifying the client's needs [10]

	Trad-itional selective	Early selective	Design & build	Construction management	Management Fee	Design & Manage
Timing						
Is early	Y		*	*	*	*	*
completion	A		*	*	*	*	*
important to	N	*					
the success of							
the project ?							
Variations						
Are variations	Y	*	*		*	*	*
to the contract	N			*			
important ?							
Complexity						
Is the building	Y	*	*		*	*	*
technically	A		*	*	*	*	*
complex or highly	N			*			
serviced ?							
Quality						
What level of	H	*	*		*	*	
quality is	A	*	*	*	*	*	*
required ?	B			*			
Price certainty						
Is a firm price	Y	*		*	*		
necessary before	N		*			*	*
the contracts are							
signed ?							
Responsibility						
Do you wish to	Y			*			*
deal with only	N	*	*		*	*	
one firm ?							
Professional						
Do you require	Y	*	*		*	*	*
direct professional	N			*			
consultant							
involvement ?							
Risk avoidance						
Do you want	Y			*			
someone to take	S	*	*		*	*	*
the risk from	N						
you ?							

Y= yes; N=no; A=average; H=high; B=basic; S=shared

- The tender prices should remain confidential at all times.
- Where errors in pricing occur, the code sets out alternative ways in which these should be dealt with. Any corrections should be confirmed in writing or initialled.
- The employer does not bind himself to accept the lowest or any tender, or take responsibility for the costs involved.
- If the tender under consideration exceeds the specified costs then addendum bills can be prepared in association with the lowest contractor's prices.
- The provisions of the code should be qualified by supplementary procedures specified in EU Directives which provide for a 'restrictive tendering procedure' in respect of public sector construction contracts above a specified value.

Co-ordinated Committee for Project Information (CCPI)

Research at the Building Research Establishment (BRE) on 50 representative building sites has shown that the biggest single cause of events which restrict the working together of site managers, designers or tradesmen is unclear or missing project information. Another significant cause is uncoordinated designs, and at times much wasted effort is directed towards searching for missing information or in reconciling the inconsistencies which arise. In an attempt to overcome these weaknesses the Co-ordinating Committee for Project Information (CCPI) [11] was formed with the task of developing a common arrangement for work sections of building works that could be used throughout the various forms of documentation. The CCPI consulted widely to ensure that the proposals were practicable and helpful, and its work has been undertaken by those working in practice who are fully aware of the different pressures involved. It is not sensible to produce incomplete, inconsistent and contradictory project information during the design process and then to expend wasted time trying to rectify the arising difficulties on site with additional time needed for dealing with claims for disruption, delays, extra costs, etc. It is preferable for complete and co-ordinated information to be prepared in order to avoid such problems. Regardless of the procurement methods used, the principles of the CCPI should be applied and adopted in practice. Procurement methods which cannot incorporate such principles are probably in themselves severely flawed in practice.

Procurement management

Clients of the construction industry rely extensively upon the advice given in respect of the most suitable method of procuring their project from inception through to completion. The advice provided must therefore be both relevant and reliable and based upon the appropriate levels of skills and expertise which are available. The need to match the client's requirements, which are sometimes vague and generally imprecise, with the capability of the industry is of vital importance if customer satisfaction is to be achieved and the image of the industry improved [12]. The procurement manager's role is therefore to:

* determine the client's requirements and objectives
* discover what is really important and what is of secondary need
* assess the viability of the project and provide advice in respect of funding, taxation, residues
* advise on the organizational structure for the project as a whole
* advise on the appointment of consultants and contractors
* manage the information and co-ordinate the whole process from inception through to completion

Summary

The traditional approach to construction has been to appoint a team of consultants to prepare a design and estimate, and to select an independent contractor. The latter would calculate the actual project's costs, develop a programme to fit within the period laid down in the contract, organize the workforce and materials deliveries and construct to the standards specified in the contract documentation. In practice, design details were sometimes inappropriate, the quality of workmanship below acceptable standards, costs above what were forecast and even in those circumstances where projects did not overrun, the completion date was later than expected in other countries [13]. In addition, too little attention has been given to future maintenance aspects and life cycle costs. The separation of the design role from construction and the different procurement practices have been the target of these obvious deficiencies. However, many clients have wished to retain the services of independent designers, believing that they will serve their needs better. They also believe in competition as a form of efficiency and have been reluctant to commit themselves to a contractor whilst costs remain imprecise. Ideally, however, the client would prefer a single point responsibility and a truly fixed price and for projects to be completed as required.

Procurement procedures remain a dynamic activity. They will continue to evolve to meet the changing and challenging needs of society and the circumstances under which the industry will find itself working. There are no standard procurement solutions, but each individual project needs to be considered independently and analysed accordingly. There is, however, a need to more carefully evaluate the procedures being recommended in order to develop good practice in procurement and to improve the image of the industry.

References

[1] Franks J. 'Further Horizons for Chartered Builders. Building Procurement, Evolving Alternative Systems'. *CIOB members handbook*, 1988

[2] Skitmore R.M. and Marsden D.E. 'Which procurement system? Towards a universal procurement selection technique' in *Construction Management and Economics*, **6**, 1988

[3] Ashworth A. *Cost Studies of Buildings*. Longmans, 1994

[4] Davis, Langdon and Everest. *Spons Architect's and Builder's Price Book*. E & F N Spon (annually)

[5] Ashworth A. and Skitmore R.M. 'Accuracy in cost estimating'. *Proceedings Ninth International Cost Engineering Congress*, Oslo, 1986

[6] Ashworth A. and Skitmore R.M. Accuracy in estimating. *Occasional Paper, No. 27*, Chartered Institute of Building. 1982

[7] *Housing and Construction Statistics*. HMSO, 1992

[8] *Building Cost and Tender Price Indices*. Davies, Langdon and Everest. London, 1996.

[9] Ashworth A. *Contractual Procedures in the Construction Industry*. Longman, 1996

[10] *Thinking about building*. Building EDC Report, 1984

[11] Azzaro D. 'Coordinated Project Information' in *Chartered Quantity Surveyor*, July 1987

[12] Morledge R. 'The effective choice of building procurement method' in *Chartered Quantity Surveyor*, July 1987

[13] NEDO. *Faster Building for Commerce*. 1988

8

Contracts

The fragmentation of functions in design coupled with the archaic contract systems ensure that integration between disciplines is at a premium.

Professor Alexander Kennaway, 1984

Introduction

An employer commissioning a construction project will expect to obtain a project that satisfies identified needs in terms of quantity, form and quality, one which is ready for occupation or use at the stipulated completion date and for an agreed price that represents value for money. Whatever procurement arrangements are selected and agreed upon, they should incorporate the following objectives:

- an agreement between the parties in respect of time, cost and quality
- a sound and logical framework for the administration of the construction process

The choice of the different arrangements will be accepted by the client on the basis of expert advice received from either a contractor or consultant, or maybe a combination of both.

Contracts in use

During the past twenty-five years changes in the methods of construction procurement have perhaps been one of the most fundamental changes that have taken place in the construction industry. Greater comparisons have been made with procurement methods used by other industries and in other countries around the world. There has also been research undertaken in order to better inform clients, consultants and contractors. The different methods available all have their advantages and disadvantages and there is no uniform solution that emerges across the industry. The choice of method depends upon a combination and importance of the following characteristics:

- type of client
- size of project
- type of project
- risk allocation
- form of contract to be used
- major objectives of the client
- status of the designer
- relationships with contractors/consultants, such as partnering (see chapter)
- sort of contract documentation required

A number of different organizations attempt to monitor procurement systems in the construction industry. The following analysis (Tables 8.1, 8.2) have been derived from 'Contracts in Use, A Survey of Building Contracts in Use during 1993'. This was prepared by the quantity surveyors, Davies, Langdon & Everest and published by the RICS in 1994. The survey is done on a biennial basis.

About 80% of the value of the work analysed in the survey used one of the JCT family of forms. The scope of the survey excluded overseas work, civil and heavy engineering, term contracts, maintenance and repairs and subcontracts that formed a part of main contracts. Work done directly by contractors were excluded from the

Table 8.1   The distribution of contracts in use [1]

	Percentage by numbe	Percentage by value
JCT80	19.36	33.55
JCT63	0.31	0.81
Minor works	26.49	1.75
With Contractor's Design	13.21	28.91
IFC84	21.98	10.40
Other JCT	0.96	4.96
Other standard forms	5.28	8.63
Employer devised	12.41	11.01

(*Source*: Davies, Langdon & Everest)

Table 8.2   Trends in methods of procurement [1]

	1984	1985	1987	1989	1991	1993
Bills of quantities	58.73	59.26	52.07	52.29	48.26	41.63
Specification	13.13	10.20	17.76	10.26	8.35	9.98
Design and build	5.06	8.05	12.16	10.87	14.78	35.70
Remeasurement	6.62	5.44	3.43	3.58	1.26	2.43
Prime cost	4.45	2.65	5.17	1.12	0.12	0.15
Management contract	12.01	14.40	9.41	14.99	7.87	6.17
Construction management	0	0	0	6.89	19.36	3.94

(*Source*: Davies, Langdon & Everest)

survey to avoid any possible double counting. This suggests that design and build represents a much larger proportion than indicated by this survey alone. This viewpoint is reinforced by the section provided later on design and build.

The use of JCT80 has declined since the last survey was carried out. This is largely due to the increase in the use of the Standard Form with Contractor's Design (JCT81). The survey suggests that 32% of 'traditional' contracts now include at least some element of contractor design. There has also been some trend towards forms specially designed by the employer, quantity surveying practices or contractors themselves.

The JCT Intermediate Form (IFC84) which was introduced for 'medium' size projects, but was often used on major schemes, continues to represent about 10% of the total value of the projects in the survey. The minor works form was used on over a quarter of the projects in the sample, although these represented only 2% of the total value of the work analysed.

It is of course difficult to derive a clear picture since the survey represents only a proportion of the total work done by consultant quantity surveyors. The effects of the continuing recession may distort the picture, since both consultants and contractors may be prepared to undertake work on less than favourable or ideal conditions. The survey represented a 15% response rate, which is typical, for anyone seeking information by way of questionnaires. This did, however, cover 3827 projects worth almost £2.5bn.

Procurement methods

From this survey, it is quite clear that bills of quantities continue to remain in use. The survey points towards their decline, at least under the traditional arrangement of a lump sum contract with a firm bill of quantities. The survey indicates a decline from 58% in 1984 to 42% in 1993. It should also be remembered that some other form of quantification for costing purposes, such as work packages, will be required with the alternative procurement methods. The survey indicates that for the large projects bills of quantities are still required.

Management contracts and construction management account for only 10% of the total value of the projects captured in the survey. The 1991 Survey revealed a dramatic fall in the use of management contracts from 14.99% to 7.87%. The latest survey shows no sign in the recovery of popularity of this form of procurement. The use of construction management has experienced an even more dramatic decline. It hardly existed at all in 1987, but by 1991 represented almost 20% of the value of projects. By 1993 this percentage had sunk to 3.94%. The considerable increase in the use of design and build methods explains some of the reasons for this fall. It might also emphasize the fact that during the recession many of the large speculative developers have ceased to be involved in new construction projects. The trends suggested by this survey indicate that whilst management contracting and construction management clearly have some role to play in the future it will be design and build that is seen as the alternative and competitor to the traditional form of procurement.

Drawings and specifications continue to be used on small works projects and these remain popular up to £100 000. There are also examples where they are used on projects with a value in excess of £1m. The construction industry in some cases has already gone full circle. Contractors being requested to prepare tenders on drawings and specifications alone then employ their own quantity surveyor to prepare quantities which they are then able to price. There are also examples recorded in these circumstances where the contractor does not price the work in detail, but prepares a tender using one of the simplified methods of estimating. The work is only priced using more detailed quantities once the contractor has succeeded in winning the contract. Also in the early days of the industrial revolution, architects used to ask trade contractors to submit individual prices for their work; this is now referred to as management contracting.

Although the general content of many of the forms of contract are similar, there are wide differences in detail and interpretation of the individual clauses and conditions. In addition to the plethora of forms which are available, their contents have become more complex. It is also more common to seek legal redress in the courts due to misunderstanding or misinterpretation. Revisions to the forms continue to take place and these, together with clause amendments and case law, have become common characteristics. Since the introduction in 1980 of the JCT80 main form of building contract, for example, there have been eight amendments. Whilst forms of contract are amended or parts redrafted for particular projects, this can be disadvantageous for the following reasons:

- omission of conditions
- inconsistency of amendments
- contradiction of terms
- bias towards one party
- unknowing transfer of risk
- mistakes
- misrepresentation
- inclusion of unenforceable conditions
- exclusion of clauses
- failure to recognize the key problems

Design and build

A special mention should be made about this form of contracting since, whilst many newer methods of contractual procurement have failed to maintain an interest, design and build continues to maintain its general share of the market. Also this method of contracting is a method that in theory at least seeks to combine the work of the designer and constructor.

This method of procurement has continued to increase in popularity, although not at the rate that had been previously forecasted. The University of Reading's Design and Build Forum [2] has recently published the most comprehensive survey of the design and build market. It is difficult to offer precise figures for comparison of the growth of design and build, since these depend upon the overall amount of work available and the general economic state of the industry. Also, in the past, when the public sector contributed more to the industry's overall output there was a bigger resistance to design and build as an option, and this therefore distorts the information.

The University of Reading survey found that design and build was 12% faster in construction speed than traditional procurement and takes 30% less time to deliver a project from the start of a design to completion. It is also 13% cheaper than traditional procurement.

Design and build projects were also more likely to finish on time. However, the technical quality of these projects is lower and worst quality occurs when novation is used. This typically accounts for about 50% of all projects. The report claims that the design and build market is static at 23%. Whilst contractors are lacking in design capability, design and build provides consistency in aesthetic quality. The report includes five key recommendations:

- Design and build must become design led.
- Contractors must employ experienced designers and have an architect on the main board.
- Repeat clients should partner with carefully selected design and build firms.
- Contractors must differentiate themselves by offering clearly branded buildings and services and offer guarantees on the quality of their workmanship.
- The Design and Build Forum will establish a design and build institute within one year to publicize the benefits of this form of procurement and to promote good practice.

New Builder (June 1995) reported that in 1992, design and build turnover for the top 25 design and build contractors fell by 2.19%. This compares with increases of 14% in 1993 and 18% in 1994. These 25 contractors include the major contractors in the industry. Design and build represented, on average, 29% of overall turnover in 1994, compared with 24% in 1993. The top five design and build contractors have more than 45% of their turnover accounted for by design and build work. On average about 10% of the larger consultants work is now described as design and build.

In the 1960s and 1970s, many people felt that design and build was only suitable for standard buildings that required little architectural design. This view has now changed to provide clients with as high quality projects as any other methods of procurement. In some cases the designer is appointed by the client to prepare outline schemes before involving the design and build contractor. This has been termed as novated design and build. Many involved feel that this puts the contractor in a disadvantaged position, providing little opportunities for innovation and improvement and hence the project's overall value for money aspects. Typically buildings requiring little design aesthetics, such as industrial premises and warehouses, were selected for design and build operations. Today, it would appear that design and build is used across the wide spectrum of project types.

Conflict

Building without conflict now seems to be unimaginable according to a recent report [3]. The main causes of conflict in the construction industry occur because of the following reasons:

Design: The growing number of parties involved who play a significant role in the project. There is a pretence that the project has been fully pre-designed, whereas in reality variations, delays and disruptions occur to the progress of the works.

Subcontracting: The vast majority of work undertaken on site is undertaken by subcontractors and their multiplicity provides conditions which require a level of co-ordination which often does not exist.

Supervision: The contractor's role is to supervise the work on site, the clerk of works is responsible for inspection and all must be done to the designer's reasonable satisfaction.

Payment: Disputes over payment are common with contractors alleging under-certification and subcontractors claiming non-payment. Should retention not always be invested with a third party?

Collateral warranties: There has been nothing which has been less welcome in the industry than the extent to which collateral warranties are required in respect of both design and construction.

Settlement of disputes

Construction contracts in the distant past consisted of a gentleman's handshake. Underlying such agreements were an essential set of values of competence, fairness, integrity and honesty. In the construction industry today a complex and onerous set of conditions that attempt to cover every possible event have been devised and in so doing create loopholes that those in the legal professions can feast upon. Precious time and resources are thus drawn away from the main purpose of getting the project built on time, to the right design and construction technology at an agreed price and quality. Three ways have been suggested for solving such disputes.

Litigation

Litigation is a dispute procedure which takes place in the courts. It involves third parties who are trained in the law, usually barristers, and a judge who is appointed by the courts. This method of solving disputes is often expensive and can be a very lengthy process before the matter is resolved. The process is frequently extended to higher courts involving additional expense and time. Also since a case needs to be properly prepared prior to the trial, a considerable amount of time can elapse between the commencement of the proceedings and the trial, as noted above. A typical action is started by the issuing of a writ. This places the matter on the official record. A copy of the writ must be served on the defendant, either by delivering it personally or by other means such as through the offices of a solicitor. The general rule is that the defendants must be made aware of the proceedings against them.

Arbitration

Arbitration is an alternative to legal action in the courts in order to settle an unresolved dispute. Most forms of construction contract incorporate conditions to refer, in the first instance, disputes between any of the parties to arbitration. The Arbitration Act came into force on the 4th of April 1979, revising the appeals procedure and reinforcing the power of the arbitrator. The advantages of arbitration include: less expensive than the courts; more speedy process; hearings held in private; time and place of the hearing can be arranged to suit the parties concerned. The arbitrator has expert technical knowledge and it can be insisted that the arbitrator visits the site if this is necessary. The arbitrator is also able to seek the opinion of the courts on a point of law if this is required.

Alternative dispute resolution (ADR)

Alternative dispute resolution is a non-adversarial technique which is aimed at resolving disputes without resorting to the traditional forms of either litigation or arbitration. The process was developed in the USA but has also been widely used elsewhere in the world. It is claimed to be less expensive, fast and effective. It is also less threatening and stressful. ADR offers the parties who are in dispute the opportunity to participate in a process that encourages them to solve their differences in the most amicable way that is possible.

Before the commencement of an ADR negotiation the parties who are in dispute should have a genuine desire to settle their differences without recourse to either litigation or arbitration. They must therefore be prepared to compromise some of their rights in order to achieve a settlement. Proceedings are non binding until a mutually agreed settlement is achieved. Either party can therefore resort to arbitration or litigation if the ADR procedure fails.

Contractual matters

Contract documents

The contract documents differ under the different forms of contract [4]. They include at least a form of contract and drawings and specification or bills of quantities (on civil engineering contracts both). They should, however, encompass at least the following information:

1   The work to be performed, which generally requires some form of drawn information. This assists both the client and the contractor by providing schematic layouts and elevations. They will also be necessary where planning permission or building regulation approval are required.
2   The quality of work required, which is normally described in the specification or the bill of quantities.
3   The contractual conditions which are the written agreement between the employer and the contractor, and should be one of the standard forms of contract which are available.
4   The costs of the finished works should, wherever possible, be predetermined before construction commences. The method of calculating these should be clearly stated either by reference to a method of measurement or cost plus definitions.
5   The construction programme, which will be prepared by the contractor to include at least start and completion dates.

Quality

The combination of sound constructional design and detailing, an appropriate specification, good standards of workmanship and efficient contractor's quality control procedures, should result in a project that achieves the desired standards in terms of quality. The contract also allows for independent inspection of the works by a client's representative and for this facility to be extended to the inspection of materials and goods off-site in the different workshops, if this is so desired. A compliance with the appropriate British Standards and Codes of Practice and a recognition of BS 5750, which deals with quality management systems, should go a long way to ensuring that the appropriate levels of quality are achieved. In addition the contractor must seek to employ competent tradesmen to carry out the work and these must be supervised by suitable site managers. Contracts also require strict compliance with the statutory regulations and allow for uncovering work for inspection or testing at any time throughout the contract period. The Building Research Establishment has shown that 50% of all building failures have their origins in design faults and 40% are due to faults of construction on site [5, 6]. An investment in quality assurance, such as ISO 9000, can therefore pay substantial long-term dividends, by reducing such failures and through savings on repair and maintenance and even legal costs.

Payment, retention and certificates

An important feature of any contract is the calculation and payment of sums of money by the client to the contractor for work which has been satisfactorily completed. The project starts with the contract sum (other than cost reimbursable contracts) which is written into the agreed contract. It is unusual, except possibly with the smallest of projects, for the whole work to be completed before any payment is made. Provision is therefore made for payment on account at the periods stated in the contract. This is normally monthly, but on very large projects this may be more frequently. An interim certificate is issued which is based upon a valuation of the works completed, and materials which have been delivered to the site. The contractor is not paid the full amount but a sum, 5–10%, is retained by the client in the possible event of default by the contractor. At the completion of the works a certificate of practical completion is issued and at the end of the defects liability period the final certificate. These have the effect of reducing the retention fund by half and in total, respectively. They also indicate that the project is technically complete and thereby eliminate possible claims for delays. The final certificate terminates the contractual responsibilities of the parties, although redress may still be taken using other legal remedies. If the project is completed in separate agreed sections or parts there are provisions for issuing a certificate of sectional completion of the works. Any interference on the part of the employer with certificates is a serious offence giving the contractor grounds for the termination of the contract and damages.

Time considerations

There are several key dates associated with a construction contract. These include the date when the work should start on site and the date when the work should be finished. It may also be required, on some projects, to allow for completion in phases or sections. If the work remains incomplete by the date for completion, without good reason on the part of the contractor, the employer is able to sue for damages. These may result from having to find alternative accommodation, to loss of production or to inconvenience. There may, of course, be good reasons why the project has been delayed, such as a lack of information from the architect or the engineer, in which case the contractor can apply for an extension of the contract period and thus be relieved of any responsibility for damages. All sorts of comparisons have been made about the length of time required for construction in Britain compared with other countries around the world. Britain fares very badly. Speed and punctuality on projects are integral to good overall performance. Slow building often signals a lack of purpose and momentum [7].

Insurances

Construction work involves risks of many kinds but especially damage to the works under construction, to neighbouring premises and personal injury or death to anyone caused by the carrying out of the works. All of the standard forms of contract require the contractor to indemnify the employer against any claims which may arise due to the contractor's negligence. The great variety of risks which have to be covered during the construction of a project and the changes in the sums involved from day to day as the work proceeds make this an extremely specialized field of insurance practice. Thus, most insurance is effected through the services of a broker with a collective scheme being arranged through the various employers' organizations.

Losses suffered by insurance companies continue to increase. Table 8.3 provides a general trend of losses which are mirrored in the construction industry. Commercial fire claims rose by 8.4% between 1990 and 1991 with domestic fire claims rising more slowly by 4.8%. The earlier decreases recorded show what could be done if the businesses and public were more careful and took full account of the prevention advice which is available. A particular problem identified by the insurance companies was arson. The positive activities of the arson prevention bureau hope that, at least, these figures will stabilize, if not decline. Theft is a major problem with increases in the number and size of claims. Thefts from businesses rose by four times the rate of inflation between 1990 and 1991. Premiums will have to rise again as result of such claims. Insurance companies identified fraud as a factor to consider. Insurance companies in 1991 were paying out over 100 claims every hour. Table 8.4 shows the increasing costs of claims attributed to subsidence damage.

Table 8.3   Annual fire wastage statistics [8]

| Year | Annual fire wastage statistics | | Year | Amount (£m) |
	Amount (£m)	Year		
1970	107	1983	566
1971	106	1984	554
1972	109	1985	450
1973	179	1986	456
1974	237	1987	461
1975	213	1988	480
1976	232	1989	600
1977	262	1990	790
1978	309	1991	780
1979	355	1992	850
1980	469	1993	646
1981	357	1994	615
1982	391	1995	700

Table 8.4   Domestic subsidence claims [8]

| Year | Domestic subsidence claims | | Year | Amount (£m) |
	Amount (£m)	Year		
1975	5	1986	95
1976	55	1987	90
1977	20	1988	90
1978	20	1989	250
1979	25	1990	505
1980	35	1991	540
1981	35	1992	259
1982	45	1993	334
1983	80	1994	125
1984	95	1995	326
1985	95		

Bond

A bond is an undertaking by a surety, such as a bank or insurance company, to make a payment to the employer if for any reason the contractor fails to complete the works. The amount of the bond is about 10% of the contract sum, depending upon the financial standing of the contractor and the experience with the type of work to be undertaken. The contractor normally arranges the bond with a firm which is acceptable to the employer, and the charges are recouped by the contractor under the terms of the contract. Local authorities are often bound by their own standing orders to require a bond when the tender sum exceeds a certain amount. The ICE form of contract states that the employer may require a bond not exceeding 10% of the contract sum but the JCT80 does not expressly provide for a bond.

Variations

Most forms of contract envisage some changes being made to the design at stages during the contract. On the JCT80 the architect must, if required, justify to the contractor the authority to issue variations. On civil engineering projects, the engineer has wider scope with the issue of variations and, in addition, does not have to justify the instruction. Variations or changes to the contract include alteration or modification of the design, the addition, omission or substitution of any work or the alteration of the kind or standard of materials or goods. A change in the circumstances in which the work is carried out also includes access and use of the site, limitations on working space, limitations on working hours and a change to the sequence of works. Variations can be given orally to the contractor but to be of any effect they must be confirmed in writing. The procedures used for valuing variations is similar on all forms of contract using bill rates, pro-rata rates, dayworks or a fair valuation which is agreed between the parties.

Fluctuations and fixed price

Contracts are either fixed price or fluctuating price contracts. The essential difference is that the former expects the contractor to have allowed for any increases in cost due to inflation over the contract period, the latter does not. Fixed price does not mean that the price cannot change, in fact it is unusual if it does not! Two different methods have been devised to calculate the increases (or decreases). The first relies on a 'formula', based upon indices, to calculate the changes in costs. This requires the contract to incorporate quantities and a more accurate assessment of interim payments. The second measures the actual changes in the costs of labour and materials, and requires a large amount of record keeping. Some fluctuating contracts only allow for changes in costs which are a direct result of changes in government legislation to be recouped. The decision to choose a fixed or fluctuating price contract depends upon the length of the contract period and the amount of inflation in the economy. When the amount of inflation is small and falling, fixed price contracts may last up to 2 years. In periods of high inflation contractors may be loathe to assess this risk beyond 12 months. On a fixed price contract, the contractor has to predict what the increases may be and allow for them within the tender. It must not be assumed that a fixed price arrangement will be necessarily more beneficial to the client.

Subcontracts

It is unusual today for a single contractor to undertake all of the construction work with his own workforce. Even in the case of minor building projects the main contractor will require the assistance of other trade and specialist firms. The spectacular growth of subcontracting occurred in the 1980s at the same time as the decline of the general contractor.

Subcontractors can be variously classified. Firms which are directly employed by the client on the construction site at the same time as the main contractors are not subcontractors since they have no agreement with the main contractor, other than perhaps access to the site to enable them to carry out their work. The forms of contract recognize them under several guises as 'persons engaged by the employer direct', 'facilities for other contractors' or 'artists and tradesmen'. Secondly, a client may choose to nominate a particular firm to undertake the specialist work that will be required. This may be to gain a greater measure of control over those who carry out the work. Such firms also have a special contractual relationship with the client. Some firms may be named in the contract documents to execute a part of the works. This approach avoids the process of nomination whilst still retaining a measure of control on the part of the contract administrator. The statutory undertakers are a special type of subcontractor since neither the employer or the main contractor have any choice over their selection. All of the remaining work is unlikely to be undertaken by the main contractor, and provision is made for the use of the contractor's own direct subcontractors. In fact, under some arrangements the main contractor may employ few workmen but manage and organize group trade subcontractors.

References

[1] 'Contracts in use' in *Chartered Quantity Surveyor*, January 1991
[2] Centre for Strategic Studies in Construction. Designing and Building a World Class Industry. University of Reading, 1996
[3] 'Building without conflict' in *Building*, November 1991
[4] Ashworth A. *Contractual Procedures in the Construction Industry*. Longmans, 1996
[5] Beard C. 'Adapt or die' in *Chartered Quantity Surveyor*, May 1988
[6] Kemp M. 'The application of quality assurance in a project management practice' in *Chartered Quantity Surveyor*, June 1987
[7] NEDO. *Faster Building for Commerce*. 1988
[8] Association of British Insurers. 1991

9

Capital and investment

Wealth creation requires investment; where the latter is small then the distribution of the former is largely academic.

Lord Scanlon
(Attributed)

Capitalization

Finance is the lifeblood of a contractor's organization. Almost regardless of the aspirations towards better building and the need for greater owner and user satisfaction, the industry is driven by costs and profits. In order to achieve this, the construction industry needs a large amount of floating capital sufficient to meet outgoings, which in the course of a year include: a wages bill of over £18bn, another £18bn on materials and components and a further £9bn for the purchase and replacement of plant and vehicles and plant hire. This is based upon a ratio of 40:40:20 for labour, materials and plant. Careful stock control systems can help to reduce the burden of materials stockpiles and storage, but the impact of off-site production can have just the opposite effect, although there is provision under the different forms of contract to allow some of this to be recouped prior to the work being completed on site.

Payment received by contractors, for contract work, each month is on the basis of an architect's or engineer's certificate, less up to 10% for retention. The retention is released to the contractor, half upon completion and the remainder at the end of the defects liability period. On larger projects the percentage retained may be smaller. Payment does not, however, reach the contractor until half way through the next accounting period. This delay, and the fact that the industry works on a 90–95% payment system, does cause difficulties to contractors, even when their accounts are up to date, and this is rarely the case. The process has a knock-on effect to subcontractors and suppliers who may not have the contractual muscle to do anything other than wait anxiously for their agreed payment to be made. However, wages must continue to be paid weekly to the majority of the operatives, although staff are paid on a monthly basis. Whilst a majority of the suppliers of materials and hirers of construction plant work on a 30 day period for payment, this is often extended as a cheap form of borrowing by contractors. The contractors' own subcontractors fare very badly and many in times of hardship have to wait twice as long as the normal accounting period. The nominated subcontractors and suppliers do a little better than the directly employed subcontractors, but the contractual terms of payment have to be checked as a regular routine by the client's quantity surveyor to avoid situations where the client may be placed in a position of having to pay twice. In some cases suppliers and subcontractors never receive payment through the all too frequent insolvency of the main contractor. Much of the plant and equipment used in the industry is now on some form of lease arrangement. This is so even where, at first glance, the site plant and equipment appear to be in the ownership of the contractor. This provides benefits in cash flow terms in that the costs are able to be spread over a number of years, taxation assessments may be reduced and in the event of liquidation all may not be lost.

At any one time about £5bn is owed to the industry for work done, and this represents a further capital commitment requirement. This amount is a minimum assuming that valuations are correct and that payments are made at the stipulated intervals. But added to this are the now large amounts of money tied up in contractual claims and legal disputes. There are also the amounts to cover the capital value of the premises, offices, workshops, storage yards, etc., of some £10bn. These sums do not include the additional sums which are tied up in the material and component manufacturers, and that invested by those professions whose livelihood rests in the industry.

Cash flow, profit, return and turnover

Cash flow

Contractors can apply a number of methods in order to maintain their cash flow, reduce their borrowing requirements and improve their overall profits. These include:

- unbalancing their tenders in favour of the early parts of the project

- over-valuation
- submitting accounts as soon as possible
- delays in the payments to subcontractors and suppliers
- reducing the levels of materials stocks

A longer-term strategy may include:

- the avoidance of work in non-profitable sectors
- the abandonment of unprofitable companies within a group
- changing the pattern of working or financing
- the purchase of new companies for: diversification
  elimination of competitors
  ownership of raw materials
  market for its own products
- investing in more mechanization

Turnover

Turnover in the construction industry in 1995 accounted for about £52.5bn [1]. This was represented by new construction work which was worth about 40% and the remainder from repairs and maintenance projects. The diminishing public sector accounted for about a third (36%) of this turnover. This continues to decline in real terms which is due in part to the effects of privatization of previously nationalized companies such as those engaged in the electricity, gas and water industries. Civil engineering now typically accounts for about 15% (£7bn) of the total output.

In 1995, the turnover of the top 50 contractors was £25bn which represents two-thirds of the total contractors' turnover [2]. The mean average turnover of these firms was £600m, although the mode was only £300m indicating that the larger firms accounted for a bigger proportion of the work. Amongst these firms Tarmac (£2.5bn), and AMEC (£1.9bn) represented the largest amounts. In 1990, the top 11 contractors, each of which had a turnover in excess of £1bn, accounted for almost £20bn; almost 50% of the total output of all contractors.

Profits

The construction industry is a high risk industry. This is evidenced by both the number of firms going out of business and by comparison with other industries. The construction industry does not normally build prototypes and this in itself adds to the risk factor. Profits often do not bear out the risks involved, and comparison with a number of indicators shows that the industry scores badly. Whilst profitability is relatively high in times of boom, these are now short-lived, unpredictable and erratic. Whilst individual contractors' profits vary considerably, they typically account for less than 5% of turnover over a period of time (see Table 9.1).

Table 9.1   Profit margins of major contractors [2]

| Contractor | Percentage profit margins | | | | |
	1989	1990	1991	1994	1995
Taylor Woodrow	8.2	8.8	5.9	3.94	4.58
Trafalgar House	9.9	9.1	5.3		
Tarmac	13.9	10.7	5.2	1.38	0.87
P&O	17.3	12.6	4.0		
Bovis				0.87	0.96
AMEC	4.7	4.6	2.9	1.46	1.34
Wimpey	8.5	6.7	2.3		
Mowlem	6.0	2.2	1.7	1.56	1.59
BICC	3.8	2.9	1.8		
Laing	5.0	4.2	1.3	2.03	1.67
Costain	7.8	3.9	0.4	8.90	−3.84
Averages	8.5	6.6	3.1	2.88	1.02

Return

This is the measure which relates profits earned to investment or capital employed. A contractor will utilize the capital employed many times per year. This may be in a ratio as high as 10:1, although it can be much lower. The stop–go nature of the industry discourages long-term investment in and by the industry. Investment varies depending upon the capital intensity of the firm, and whether, for instance, the firm has a large land bank or uses property as a source of income.

Investment

Investment is necessary for wealth creation (Figure 9.1). The products of the construction industry, i.e. buildings and other structures, have long been seen as a good source of long-term investment. In periods of rapidly rising property prices and when the demand for accommodation exceeds supply, the return to investors, even in the short term,

Figure 9.1   *Investment is necessary if wealth is to be created*

has often been high. Sometimes it has been artificial and only the wise investor has been able to see that on occasions this has been short-lived. Whilst the majority of other new goods and products depreciate after purchase, buildings and other structures have, in the past, always shown an increase in their values. The phenomena of the early 1990s experience, of declining property and rental values, has been unprecedented in recent times, resulting in mortgage lending on some properties being unsecured. However, pension funds, insurance companies and private investors continue to see property as a good hedge to their savings above the level of inflation. This has, in some cases, reaped financial rewards over a matter of a few years. Over the longer term, investment in property shows a real mark-up above many other commodities. However, the future can never be certain. The recession of the early 1990s saw a collapse in property prices throughout the country by as much as 50% in some areas. Some reports also suggest that it may be the end of the decade before property values reach the high levels that were achieved in 1989. This is bad news for investors in property and even worse news for a depressed construction industry.

It is not possible to provide a realistic total value of all property in Britain, due to the vagaries in the property market. The value of domestic housing is estimated to be worth £800bn, of which over 50% is accounted for in the south east of England. The average value of a house in Britain in the early part of 1992 was approximately £67 270. This varied from £48 720 in the north of England to £91 581 in Greater London, according to the Household Mortgage Corporation. By 1996 this had risen to over £100 000 in London. These figures reflect the slump in house prices which occurred in the early 1990s, and is also a wider indicator of the general slump in property prices. The HMC expects house prices to increase over the next four years by an average of almost 35%, increasing most in Greater London by almost 50% with Scotland showing the least increase (11%). By comparison, the most expensive one bedroom flat (in London) was for sale at £3.275m in 1990. Much of the price differentials are represented by the differences in their respective land prices. The maxim that location, location and location are the three most important factors which affect property values is evident from the house price market.

During recent years the construction industry has been engaged on several major and innovative projects such as the Channel Tunnel, the Canary Wharf office development, offices at Broadgate in the city of London, electrification of the east coast mainline for British rail, Sheffield student games projects, and the Meadowhall shopping centre development, also in Sheffield. These along with other schemes under construction and development both at home and overseas, represent considerable investment in the British construction industry.

Fixed assets including plant

Table 9.2 shows the gross capital stock, i.e. buildings, plant and machinery, vehicles, ships and aircraft. These are based on 1985 replacement costs for the construction industry, the manufacturing industry and the financial services sector and are for the United Kingdom as a whole. Also shown are the ratios of the capitalizations based upon the respective gross domestic products.

The ownership of fixed assets by the construction industry is low compared to the other industries shown. For construction, about 60% of the fixed assets are plant and machinery. In terms of the gross domestic product produced by each industry, the fixed assets are strikingly low in construction and confirm that it is labour intensive. It is also a reflection of the insubstantial nature of the construction industry. During the last decade, the capitalization ratio for

Table 9.2 Gross capital stock at the 1985 replacement cost and the capitalization ratios, i.e. the ratio of capital stock to gross domestic product for each industry [3]

	1980	1981	1982	1983	1984	1985	1986	1987	1988	1989
					(£,000m)					
Construction	15.6	15.6	15.8	15.9	15.8	15.9	15.9	15.9	16.3	16.6
Capitalization ratio	0.96	1.05	0.99	0.93	0.88	0.89	0.84	0.78	0.74	0.71
Manufacturing	252.9	255.2	257.2	258.9	261.4	264.9	267.4	270.1	273.4	277.8
Capitalization ratio	3.55	3.81	3.97	3.75	3.65	3.61	3.59	3.45	3.26	3.18
Financial services	71.4	75.7	80.3	84.9	90.1	96.0	102.7	111.9	123.7	138.2
Capitalization ratio	2.18	2.22	2.21	2.18	2.14	2.14	2.06	2.04	2.07	2.22
Total: Industry	1377.6	1403.6	1430.8	1459.4	1490.8	1524.2	1555.9	1593.2	1637.4	1687.6
Capitalization ratio	4.95	5.12	5.12	5.05	5.02	4.97	4.91	4.80	4.73	4.77

construction has reduced considerably. This is due to an overall increase in gross domestic product linked with little or no increase in the fixed assets. Although it is suggested that the use of the fixed assets has become more efficient and effective during the past decade, this is not entirely the case. A good deal of the expansion in construction in recent years has been in areas in which relatively little use is made of plant, for example, in new housing and repair work. In addition leased assets are not included although these constitute a comparatively small proportion of the total gross capital stock.

Table 9.3 shows the annual expenditure during the past decade on fixed assets revalued at 1985 prices. This too covers a number of industries and includes the capitalization ratios based on the annual expenditure and the gross domestic product for each industry listed.

Table 9.3 shows the same trends as that of Table 9.2. Compared to other industries, construction has a relatively low investment in fixed assets and, in terms of the gross domestic product, this investment has tended to reduce during the past decade. The construction industry as a whole has decided to avoid working practices that demand a substantial investment in plant and machinery. Contractors have subcontracted a larger proportion of their work and reduced their commitment to fixed capital. By this means they have no responsibility to replace labour by enhanced mechanization. Construction projects are also discrete entities and plant is required only for limited periods as are the trained personnel that operate the plant. In order to cater for the needs of the construction industry, in part, by providing plant and trained operatives as required, a separate plant hire industry is available. This accounts for about 50% of the contractors' requirements and, apart from difficulties during severe recessions, the system works fairly well for all of the parties involved. The contractor has less fixed capital locked into production and is able to have access to plant and trained personnel without maintenance and breakdown problems. Provided that there is an appropriate level of construction work, the plant hire industry can supply the required range of plant to the successful tenderers. Nonetheless, in terms of the economic operation of the industry, the separation of much of the productive capacity of the industry from its production process acts as a brake on innovation. It is true, of course, that items of plant are continuing to be improved, but this tends to be dependent on the plant manufacturer rather than through innovation by the contractor. The civil engineering sector of the industry is the most mechanical plant intensive where its value can typically represent as much as 25%, and even more on some projects which require the need for major mechanization. However, on construction as a whole, which includes the repair

Table 9.3   Annual expenditure on fixed assets at the 1985 replacement cost, and the capitalization ratio [3]

	1980	1981	1982	1983	1984	1985	1986	1987	1988	1989
						(£m)				
Construction	630	567	653	691	609	626	582	687	999	922
Capitalization ratio	0.039	0.038	0.041	0.041	0.034	0.035	0.031	0.034	0.045	0.039
Manufacturing	9 781	7 672	7 482	7 410	8 823	10 118	9 423	10 048	11 198	12 386
Capitalization ratio	0.138	0.115	0.112	0.107	0.123	0.138	0.127	0.128	0.134	0.142
Financial services[a]	4 349	4 756	5 239	5 417	6 203	7 133	7 985	10 819	13 786	16 664
Capitalization ratio	0.133	0.140	0.145	0.139	0.147	0.159	0.161	0.198	0.231	0.258
Total: Industry	53 416	48 298	50 915	53 476	58 034	60 353	61 813	67 753	76 648	81 845
Capitalization ratio	0.192	0.176	0.182	0.185	0.196	0.197	0.195	0.203	0.204	0.231

([a] includes leased assets)

and maintainance sector, as little as 5% of the cost of a project may be spent on plant and equipment [4].

Ratio analysis

Ratio analysis is a tool with which to interpret financial and other information. Comparisons can be made with equivalent ratios from previous years and with those of the industry in which the company is positioned. If the information is extracted on a regular basis then an individual firm's management can also have some indication of the trends which exist, and be able to take appropriate corrective action in advance of possible difficulties arising. The ratios lead to asking the right questions rather than providing conclusive answers.

Liquidity

The first concern of any manager is to ensure the short-term survival of the company. Is the company able to meet its immediate obliga-tions? Commentators prefer to see a company with more current assets than current liabilities, and with a ratio of at least 2:1. An alter-native to this is the quick ratio. This provides a more rigorous mea-sure of the company's ability to meet its short term obligations by removing the value of stock, some of which might represent unsaleable lines. A general contractor's stock would largely be repre-sented by materials and components. These could be worth anything up to 20% of current assets but, due to the decline in general contract-ing in favour of labour-only gangs, this might represent a much lower amount. Liquid assets represent the cash available together with the outstanding debts. Because of the nature of the construction industry, a company having work in progress or completed cannot be said to have a current asset readily convertible to cash. Accountants tend to look for an acid test ratio of 1:1.

Capital structure

The net assets of a company can be financed by a mixture of owners' equity and long-term debt. Gearing ratios analyse this mixture by measuring the contributions of shareholders against the funds pro-vided by the lenders of the loan capital. The profit and loss account provides another useful angle on the capital structure. Is there a healthy margin of safety in the profits to meet the fixed interest pay-ments in the long-term debt? An over-geared company may show signs of running out of profit to pay this fixed burden. To be sure that their dividend is safe, shareholders will want profits compared with the dividends payable.

Activity and efficiency

The ratios showing stock turnover and average collection period help managers and outsiders to judge how effectively a company manages its assets. The figure of sales is compared with the investment in various assets. The retail sector, for example, tends to show a rapid stock turnover emphasized by the phrase, 'past its sell-by date'; manufacturing companies show a much slower turnover. Construction is somewhere in between. The collection period can be as little as a few days. In business in general it is typically 30 days but can be as high as 60 days in the manufacturing industry. In the construction industry the situation is aggravated due to the application and variations in retention sums. Similarly, for cash flow purposes, managers should aim to extend the period of credit taken to pay suppliers. However, too long a period will lead to poor trade relations with suppliers generally and may present a picture of a poor financial position in a company.

Profitability

This profit margin ratio shows managements' use of the resources under its control. Extraordinary items should be excluded for comparative purposes. These do not represent the normal operating profits. Profits are closely related to the assets in a company. Some analysts will calculate the return on specific assets, e.g. the inventory. If the quoted company fails to earn an acceptable return, the share price will fall and prejudice chances of securing additional capital or a long-term debt on beneficial terms. These ratios need to be examined in the context of the gut feeling of the firm, and understanding of the general state of the construction industry, national and, where appropriate, international economics. When there are new contracts to be won the possible over-valuation or the losses encountered on a project may easily be absorbed or offset the possible increase in profits. When recession occurs and tender margins reduce such possible losses may be too large to allow the firm to continue trading [5].

Finance

Short-term finance

There are several ways in which firms in the construction industry supplement their long-term finance for short periods of time. The usual way is by borrowing from a bank on a short loan or overdraft facility. The interest rate charged will often be at least 1–2% above the bank rate. This is the rate at which the Bank of England will lend money for short periods to high security customers. A large part of the construction industry does not come within this category and the percentage charged is usually much higher. The lender is often more concerned with the security aspect than the interest earned; but both may be at risk. Such capital may be required to smooth out the strains on cash flow resulting from over-trading or fluctuations in the demand of work. The repayment of short-term loans is generally within 2 years. A firm may further supplement its short-term needs by:

1 *Trade credit.*   Delaying the payment of creditors beyond the normal terms. The discounts offered for prompt payment may be foregone in order to improve the cash flow.
2 *Subcontracting.*   This helps to relieve the initial outlay on goods and materials.
3 *Hire purchasing.*   The company can enjoy the full benefits of use by making deferred payments.
4 *Leasing.*   This has now become a more common procedure because of taxation benefits.
5 *Stock control.*   Reducing the levels of goods and materials to meet only immediate needs.
6 *Venture capital.*

Medium-term finance

Medium-term capital is that which is required for periods of 5–10 years. Approximately 40% of all business loans are taken out for this period of time. Most of these loans are secured against the assets of the business or are guaranteed by their owners.

Long-term finance

Long-term capital, which usually means capital for 10 years or more, is that provided to allow a business to purchase office and workshop space or for plant and equipment which will have a considerable lifespan. It may also be required to finance takeovers. If the business is sound, and it will probably need to be, then it may be better to attempt to raise such finance through a share issue rather than to take on long-term debts. Banks may be reluctant to lend money where there is a lack of confidence about the long-term prospects for growth. Table 9.4 indicates the amount of lending from UK banks to the

Table 9.4   Bank lending to construction [6]

	Amounts outstanding (£bn)		
	To Construction	Total to UK residents	% Constructions
1985	5.1	168	3.03
1986	5.7	203	2.81
1987	7.2	243	2.94
1988	10.8	305	3.54
1989	15.1	413	3.66
1990	17.3	452	3.82
1991	16.4	477	3.43
1992	15.5	468	3.31
1993	13.1	479	2.73
1994	11.5	487	2.36

construction industry which has consistently been between 2% and 4% of the total. Long-term capital may be a combination of:

- retained profits
- clearing bank loans
- merchant bank loans
- industrial and commercial finance corporation
- shares
- debentures
- government grants and loans
- sale and lease back

Contractor's interest cover

The High Street's textbook cover is that operating profit should cover interest charges by a ratio of 3:1. In demonstrating an ability to repay a debt, the ratio is the clearest indication of a company's health. It focuses on operating profit ignoring the distorting effect of provisions in a company's figures. Before the recession of the 1990s most companies could demonstrate this requirement. Also, according to banking sources, construction covenants are now reviewed about every 6 months compared with arrangements which were set over a 3-year period in the late 1980s. Many of these contractors (Table 9.5) had insufficient cover in 1991. An exception was Laings, which, it was reported, made sufficient gain from their bank balances to cover the

Table 9.5   Contractors Interest cover [7]

	Ratios			
	1989	1991	1994	1995
Costain	4.25:1	1.70:1	—	—
Tarmac	6.50:1	2.19:1	4.30:1	3.50:1
Wimpey	4.45:1	2.25:1	—	—
Higgs & Hill	13.88:1	2.59:1	—	—
P & O	6.17:1	2.70:1	—	—
Bovis	6.17:1	2.27:1	—	—
Mowlem	5.46:1	5.18:1	3.42:1	4.32:1
Laing	9.98:1	–	no interest charges	
AMEC	17.20:1	–	5.20:1	4.04:1
Taylor-Woodrow	—	—	5.31:1	4.85:1

interests on their borrowings. Costain, on the other hand, in the mid 1990s made losses.

Share capital

Shares are the equity capital of a company, with the shareholders entitled to the residual profits in the company and to voting rights. In 1988, the pattern of share holding was as shown in Table 9.6.

The market capital of *The Times* 100 top companies, across all industries, compiled by Extel Financial is worth over £400 000m (1992). The top 10 companies now account for over £140 000m or 35% of this listing. Construction activity, which also includes the material producers and component manufacturers amongst the top 100, is very small representing only about 1–2% of the market capital. The

Table 9.6   Typical pattern of share holding

Pension funds	32%
Insurance companies	25%
Individuals	20%
Unit trusts	6%
Overseas	5%
Investment trusts	4%
Others	5%
Government	3%

Table 9.7   Total business failures for 1980/1991 [8]

	Liquidations	Bankruptcies	Total
1980	6 814	3 837	10 651
1981	8 227	4 976	13 203
1982	11 131	5 436	16 567
1983	12 466	6 821	19 287
1984	13 647	8 035	21 682
1985	14 363	6 580	20 943
1986	13 689	6 991	20 680
1987	10 644	6 761	17 405
1988	9 276	7 286	16 652
1989	10 197	7 966	18 163
1990	13 936	14 999	28 935
1991	20 736	27 038	47 777
1992	22 938	39 829	62 767
1993	19 672	36 061	55 733
1994	16 335	27 263	43 598
1995	17 280	24 023	41 303

share value of those companies listed in the construction industry, fluctuates both in respect of the industry's own performance and also the general level of the company's activities.

Sharewatch, a feature of *Building* magazine, charts the value of selected shares in the industry on a weekly basis, and compares these with shares recorded by the *Financial Times* indices. The share markets recorded include the FT industrial ordinary, contractors, building materials producers, house builders' share, which is a weighted index of the top 15 UK house builders and the *Building* 50, which is a weighted index of the share price of the top 50 UK quoted contractors.

Liquidation and bankruptcy

Limited liability companies cannot go bankrupt. If they are insolvent then they go into liquidation. Bankruptcy relates only to individuals and therefore there is no such thing as a bankrupt company. The annual report on bankruptcies from the Department of Trade and Industry is often greeted with a mixture of sensationalism and 'as expected' as far as the construction industry is concerned. However, the industry does contribute disproportionately to the figures. A company becomes insolvent when the value of everything that it owns comes to less than the value of its debts. The company may enter voluntary liquidation or, more commonly, have been forced in this direction by a single creditor. The single creditor is often a financier who refuses to extend a loan or chooses to call in a debt. Even a profitable firm may suffer in this way because of a shortage of cash to pay bills. It thus suffers from a cash-flow crisis.

The 1991 figures (see Table 9.7 and Figure 9.2) are the highest ever recorded. They represent a rate of collapse of 995 companies a week or almost 200 for every working day. For companies to weather the business climate it is essential that they are careful with whom they do business, employ sound financial disciplines and verify and monitor companies. In times of recession new, as well as established companies are going to the wall and more companies do not pay their bills on time. A way out of the vicious circle is to introduce legislation to improve payment performance, otherwise the long-term situation will get worse. Over 20% of the business failures occurred in the south east of England in 1991, plus a further 16% in the London area.

Insolvencies in the construction industry more than doubled between 1990 and 1992, from an already high base (see Table 9.8 and Figure 9.3). These exclude statistics for consultants and building manufacturer producers. The kind of firm that goes into bankruptcy or liquidation ranges from the one-man business, subcontractors, suppliers, general contractors through to, on occasion, the national

Table 9.8   Number of administrative receiverships and administration orders in the construction and property industries [9]

		Building	Property
1989	3Q	70	5
	4Q	90	10
1990	1Q	180	20
	2Q	130	15
	3Q	140	30
	4Q	220	40
1991	1Q	260	50
	2Q	280	60
	3Q	210	60
	4Q	230	80

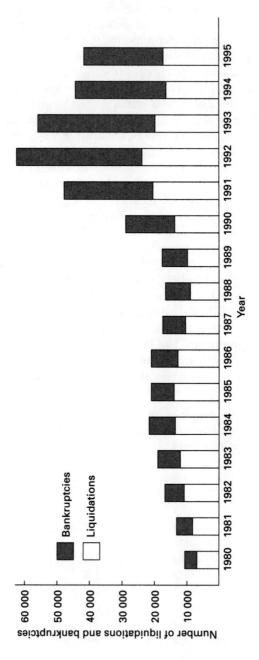

Figure 9.2 *Liquidations and bankruptcies [8] (all industries)*

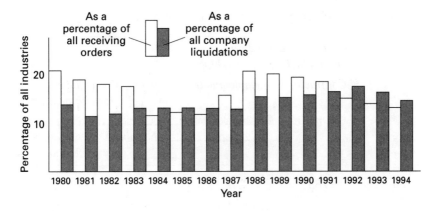

Figure 9.3   *Insolvency in the construction industry [1]*

contractor. When this happens a knock-on effect is created and other firms teetering on the edge often go with them.

Table 9.8 partially disguises the difference in sizes between the two sectors and the actual size of the failures. The relatively small numbers of property companies recorded is balanced by the fact that they are on average much larger than the average company in the building category. Tables 9.9 and 9.10 indicate that more construction firms went out of business than ever before. In 1991, 990 building related firms and 266 property developers ceased trading. Company failures in the construction industry represent about 15% of all company failures. This is a much higher figure bearing in mind that construction represents only about 6% of total companies.

Table 9.9   Construction insolvencies by sector 1991 [9]

	Number	*Percentage*
General construction/demolition	739	59
Domestic repair and improvement	56	5
Decorating and small works	69	5
Building repairs	38	3
Electrical and plumbing	88	7
Property	266	21
Total	1256	100%

Table 9.10   Insolvency in the construction industry [1]

| | Receiving orders administered, individuals & partnerships | | Compulsory and creditors voluntary liquidations | |
| | Contractors/self employment | | Construction companies | |
	Number	Percentage of all orders	Number	% of all liquidations
1980	783	21.1	949	13.8
1981	972	19.1	990	11.5
1982	968	18.4	1422	11.8
1983	1180	17.8	1776	13.2
1984	901	11.6	1831	13.3
1985	788	12.3	1975	13.3
1986	801	11.8	1914	13.3
1987	1123	16.0	1490	13.0
1988	1590	20.6	1471	15.6
1989	1652	20.3	1638	15.1
1990	2348	19.5	2445	16.2
1991	3812	16.8	3373	15.5
1992	4692	14.6	3830	15.7
1993	4361	14.1	3189	15.4
1994	3362	13.1	2401	14.4

The role of the liquidator

The commonly held view of receivers and their advisers is one of an undertaker to perform the last rites for an ailing construction firm. The role adopted by the receivers is to preserve and salvage any or all the parts of the firm. This is based upon the fact that where insolvency occurs all those involved are losers; there are no winners. The types and values of assets which can be recovered vary considerably depending upon whether the company is a contractor or developer, manufacturer or even a consultant. Contractors have become wise enough to separate the company's activities into groups such as plant hire, house building, general contracting, etc., so that if hard times occur then the whole business might not be lost.

When insolvency becomes a possibility then speed becomes all important. The whole assets of the contractor must be maintained and not traded against favours elsewhere. In the construction industry when the receivers are appointed, a firm of specialist quantity survey-ors arrives at the same time. For each contract there are three choices available:

1 Completion of the project. This is desirable and the more advanced a project is towards completion then there is a better likelihood of this occurring.
2 Abandon it, where the cost to complete is less than the value of the project. This is the least desirable option because there are usually assets still tied up in it.
3 Sell it to a third party.

In each of the above cases the quantity surveyor calculates the amounts of work done, the materials on site, what is owed to subcon-tractors, suppliers and other creditors, amounts outstanding in reten-tion, etc. In about 75% of receiverships, projects are either completed or sold to other firms. For example, a contractor could be 60% of the way through a £3m project, but of the £1.8m work completed the con-tractor might only have received £1.5m owing to work being com-pleted since the interim valuation and outstanding retention sums. Coupled with this there is a good possibility that future profits may exist in the remaining work which needs to be carried out. However, life is never down to simple arithmetic and other factors will need to be taken into account such as the amounts owing to the different creditors, the lowering of the site morale and motivation and the fact that the project might have been financially front-loaded.

Current analysis

The construction industry went through a difficult period during the early 1990s, and was showing few signs of improvement. Many of the large national contractors consider that it will be well into the late-1990s before any real upturn occurs on the building sites. Some prop-erty analysts are suggesting that upturn may not occur until 1997; almost at the time of a general election.

The gloom in the industry is justified. In 1991 output fell by a fur-ther 10%, employment by 14% and the number of bankrupt builders increased by 50%. The industry had a turnover of £45bn in 1990 and £42bn in 1991. Many have judged this to be the worst recession in memory. At least 250 000 of the 1.5 million workforce have been sacked. Behind these grim statistics are scattered many wrecked corporate plans, hundreds of dissolved partnerships, and tens of thousands of disrupted careers. The NEDO forecast in the short term also makes grim reading. It expects output to fall still further in the immediate future. The commercial sector, for example, declined by a further 20% between 1991 and 1992 and fell by 27% in 1992 and 8% in 1993. Even the industry's normally reliable repair and maintenance sector is in decline. Whilst it currently represents 50% of all output, this fell by 11% in 1991. The civil engineering sector, which grew by 11% in 1990, was flat in 1991 and fell by 3% in 1992. Some of the figures may be misleading since they are all based on cost and tender sums which were below the absolute minimum. In road building, for example, tenders are being let up to 30% below the forecasted costs.

The glimmer of hope is at the moment overseas, with large projects being let to British contractors in Saudi Arabia, and over 1000 firms involved with the Euro Disney project near Paris. Europe is expected to grow as a market generally and for British firms in particular. Architects, engineers and quantity surveyors are opening up small offices in many European cities and forming joint partnerships in the EU and the former eastern European countries. Government is also able to pump prime through public spending, on the time-held assumption that it is (always?) the construction industry which signals the ending of a recession. Government is also able to reduce interest rates. Eventually something happens to create confidence for the cycle to start again.

Cut-throat competition, wafer-thin profit margins and high restructuring and reorganization costs have been the hallmark of the early 1990s. The promise of the end of the recession 'tomorrow' is still the cry heard throughout the industry. 1994 looked modestly promising but these were reversed through disappointing results of most firms in 1995. The move towards the Millennium and the fact that the industry is becoming more realistic about workloads are reasons to be optimistic about the future. Many argue that the recession must end some time. Just when, is the question most frequently asked question by those employed in the industry and by those who commentate on it.

Taxation

Taxation affects the construction industry in much the same way as it affects an individual, through direct and indirect recovery. The application and rates of recovery change at each new Finance Act which follows the budget issued by the Chancellor of the Exchequer. Matters which are of interest to the construction industry are as follows.

Value Added Tax

The current legislation on VAT was initially introduced in 1972, to replace the purchase or sales tax which had been levied since 1940. VAT has undergone many changes both as a result of a desire to close loopholes in the application of the tax laws and also to adjust the rates and the items on which it is to be applied. There are at the present time essentially two rates of tax; standard rate which is currently 17.5%, or a zero rate. The process of levying VAT has two aspects. At each stage of production, VAT is calculated at the input stage and the output stage. When a contractor purchases materials from a supplier these will include VAT as an input tax to the contractor. It is collected on behalf of the Customs and Excise by the supplier. When the materials have been incorporated within the building and the building owner has been invoiced, this will include VAT as an output tax to the contractor. The contractor pays to the Customs and Excise the difference between the output tax and the input tax. New residential buildings are, for example, eligible for zero rating within the current VAT structure.

Statutory tax deduction scheme

The purpose of this legislation, which was introduced in 1975, was to deal with the problems of tax evasion by subcontractors' particularly the labour only or 'lump' subcontractors. The main contractor collects tax on behalf of the Inland Revenue from those subcontractors who do not hold such a certificate. Those subcontractors who hold a 714 certificate are eligible for full payment on the basis that they deal with the Inland Revenue direct. There are currently about 625 000 [10] subcontractors who hold 714 tax exemption certificates. However, a proposal by the Inland Revenue to create a turnover limit of £25 000 would exclude 525 000 subcontractors from holding a certificate which allows employers to make gross payments with no deduction of tax.

Capital allowances

Some types of capital works items, e.g. some types of buildings and equipment, are eligible for having their costs offset against company profits. The items and the various percentages, which may be claimed as an initial sum or allowances for future years, are determined by way of the Chancellor of the Exchequer's budget and the ensuing Finance Act [11]. Although the Finance Act defines the criteria upon which capital expenditure qualifies for relief, subsequent legal case laws may also be used to interpret the Act. The writing down allowances are given in consideration of the depreciation of the asset. However, in practice there may be little relationship between the allowance, the accounting amount for depreciation shown in the company's books and the actual depreciation of the asset. Assets are often more quickly written down for tax than for accounting purposes. The underlying principle of government is to encourage companies to invest in order to improve the country's performance and competitiveness. In accordance with these principles, plant and machinery are usually allocated preferential writing down rates. The initial project needs a careful analysis to separate out those items which are able to generate this initial and ongoing cost advantage. The Inland Revenue provides for three separate items; initial allowances, writing down allowances and balancing charges.

Financial assistance

In order to encourage certain types of development to take place in certain areas, government has traditionally provided forms of financial incentive packages through regional aid. These have encouraged industrial or commercial buildings to be constructed or provided local authorities with special funding to develop the infrastructure necessary to encourage companies into an area. The European Social Fund (ESF) has also provided money to designated areas in order to soften the unemployment difficulties in areas affected by European Commission policies, e.g. the restructuring of the steel industry. Aid is given towards new capital projects which include buildings, machinery or equipment. There is a wide range of different agencies offering advice and financial assistance.

References

[1] Department of the Environment. *Housing and Construction Statistics.* HMSO
[2] 'Top fifty firms', survey in *Building*, 1991
[3] Central Statistical Office, *Annual abstract of statistics.* HMSO. London, 1990
[4] National Economic Development Office. *How flexible is construction? A study of resources and participants in the construction process.* HMSO. 1978
[5] Ashworth, A. 'How to cut your losses' in *Building*, February 1978
[6] Bank of England Annual Report

[7] Duffy, A. 'High bank charges cramp contractors' in *New Builder*,
     April 1992
[8] Dunn & Bradstreet International.
[9] Stewart, A. 'Sizing up trouble' in *Building*, January 1992
[10] Yorke, T. A. 'Dead Cert' in *Building*, April 1991
[11] Inland Revenue *Capital Tax Allowances.* (annually)

10

The professions

Fragmentation of activity with the inevitable increases in interfaces between one discipline and another, must cause inefficiency and loss of control. It has an impact upon quality. Too many competing professions make the task of management harder.

Sir Christopher Foster, 1991

Introduction

Designing, costing, forecasting, planning, organizing, motivating, controlling and co-ordinating are some of the roles of the professions involved in managing construction, whether it be new build, refurbishment or maintenance [1]. These activities also include research and developing and improving standards.

Definition of a profession

It is easy to identify those occupations in society which constitute a profession but much less easy to explain why some groups are included and others are not. The mere fact that an individual may belong to a professional society does not automatically mean that he or she is a professional. Not all white collar workers are, for example, members of a profession. There are four essential attributes which generally account for professional status.

1 A body of systematic knowledge which can be applied to a variety of problems. For example, structural engineers have a body of knowledge which they can apply to projects to determine the structural behaviour of a part of a structure.

2 Professionalism involves a concern for the interests of the community at large rather than just self-interest. Thus, the primary motivation of a professional is in the interests of the client rather than personal gain. On some occasions it may be necessary to advise a client not to go ahead with a project and thus lose a commission rather than to develop a project which is not worthwhile.
3 The behaviour of professionals is strictly controlled by a code of conduct or ethics which are maintained and updated by a professional association and understood by those in training as a requirement of qualification as a full professional member. If members break the code, then the association can impose restrictions upon their activities or ban them from practising using the benefits of membership. In the case of an architect, expulsion from ARCUK means that the individual can no longer be described as an architect.
4 Professionals receive high rewards in terms of earnings and status within society. These represent the symbols of their achievements, reflect their contribution to society and the view held by society.

Higher and lower professionals

In terms of their market situation, the professionals can be divided into two groups; the higher and lower professionals. The higher professionals include the chartered construction professionals as well as accountants, lawyers, doctors, university lecturers, etc. The lower professionals include school teachers, nurses, social workers, librarians, etc. There are significant earning differences between the two groups as indicated in Table 10.1.

Measured in terms of earnings, the higher professionals as a group do much better than the other groups and in addition enjoy better fringe benefits such as company cars, health care schemes, etc.

Table 10.1 Relative earnings of the main professional groups [2]

Occupational group	Average pay as a percentage for all men						
	1914	1922	1935	1955	1960	1970	1978
Higher professional	230	206	220	191	195	155	159
Managers/administrators	140	169	153	183	177	180	154
Lower professionals	109	113	107	75	81	100	104
Skilled manual	74	64	68	77	76	76	83
Unskilled	44	45	45	54	51	61	65

The market situation of the lower professionals is not substantially superior to those of skilled manual workers. However, they do enjoy better conditions of service in respect of security of employment, wider promotion opportunities, annual salary increments and some fringe benefits.

Growth

The professions have been one of the fastest growing sectors of the occupational structure in Britain. At the turn of the century they represented about 4% of the employed population. In the early 1970s this had risen to over 11% and the trend in their growth has continued with an acceleration as the service sectors increased their importance during the 1980s and manufacturing either became more mechanized or generally declined. The temporary lull in the expansion of the professions, due to the recession of the 1990s, has caused much discussion on their benefits to society. A similar trend of comparable groups is evident in all western capitalist societies. Several reasons are given for the rapid growth of the professions, such as an increasing complexity of commerce and industry, the need for more scientific and technical knowledge and an improved desire for greater accountability.

The built environment professions

The built environment professions [3] in Britain are many and varied and represent a distinctiveness about the industry and a matter for much debate. It is argued that the difficulties which arise in the industry are due, at least in part, to the many different professional groups which are involved. There are others who suggest that the services which the British construction industry provides have now become so specialized that one or two different professional groups would be inadequate to cope with the complexities of the British construction process. In this respect, Britain is out of step with the rest of the world. However, there is no standardization of practice and considerable differences in practice exist even across mainland Europe. Worldwide, different practices have been developed which are now really a matter for history.

Architects

Traditionally the architect was the first point of contact with the client who was contemplating construction. Historically, therefore, the architect became the leader of the building team. However, this position has been eroded as clients have opted for different procurement relationships depending upon their own particular needs, such as design and build, and other similar 'one-stop shopping' arrangements. These have, over the past few decades, resulted in clients entering into single arrangements with contractors, and this has moved the focal point of the project away from the designers. During this century the role of the architect in Britain has become more narrowly focused, as other professions have impinged upon their work.

The architect's function is to determine a proper arrangement of space within the building; its shape, form, type of construction and materials to be used, environmental requirements and aesthetic considerations. This is all now done within the concept of a whole project life approach. The architect prepares the design, obtains planning permission and building regulation approval, prepares detailed drawings and specifications, advises on a contractor and inspects the work under construction. The architect's role is described in the form of a contract between the client and the contractor, such as the JCT80. The scope of activities undertaken can be subdivided between pre-contract (design) and post-contract (inspection), although it is common for an entire service to be provided. The pre-contract phase entails three facets; architectural design, construction detailing and administration. This latter aspect involves integrating the work of the various job architects, technicians and other consultants and ensuring that the information is available for the different stages of the project. The responsibilities are established in line with trade custom and practice. During the post-contract phase the architect's duties are largely one of inspection of the works to ensure that they comply with the contract. The architect does not supervise the contractor on site. The amount of inspection which is required varies with the type of project. Refurbishment schemes, for example, are likely to require more frequent inspection than new buildings. During the post-contract phase the architect issues instructions and certificates to the contractor.

The Royal Institute of British Architects

The RIBA is the main professional body for architects although a number of smaller and non-chartered bodies also include architects amongst their membership. In Scotland, a similar body exists known as the Royal Incorporation of Architects in Scotland (RIAS). Under the Architects Registration Act 1938 it is illegal for anyone to carry on the business as architect unless they are registered with the Architect's Registration Council (ARCUK) established under an Act of 1931. Registration involves appropriate training and education as evidenced by the possession of qualifications set out in the Act or approved by the Council. However, this does not prohibit the carrying out of architectural work, such as design. In 1995 there were 31 700 members of the RIBA, of whom 27 227 were corporate. In addition there are a few thousand members in the RIAS. There were 30 500 architects registered with ARCUK in 1996. Of these, the largest percentage are employed in private practice, many in the capacity of principals. Other architects are employed by central and local government and a small number by builders and development companies.

Surveyors

There are a number of different types of surveyor. The quantity surveyor has developed from the role and function of a measurer to that of cost consultant. The emphasis of their work has moved from that which was solely concerned with accounting functions, to those involving the financial forecasting of construction projects. The quantity surveyor's role is threefold. Firstly as a cost consultant at the strategic and conceptual phase of pre-design, both on an initial cost and whole lifetime basis. Secondly, in preparing tendering and contractual documentation for use by general and specialist contractors. Thirdly, in an accounting role during the construction phase where reports are made for interim payments, financial progress and control, the adjudication of contractual claims and the preparation and agreement of the final project expenditure. The quantity surveyor is employed across the whole spectrum of projects on buildings, civil engineering and petrochemical works, and on behalf of the client and the contractor.

Traditionally the building surveyor's role was in assisting other professional colleagues on the maintenance and repair of buildings and in the preparation of survey reports for the prospective purchasers and property users. This is the construction industry's most rapidly expanding profession. Some of this is due to the changing emphasis of construction and on the need to adequately maintain and repair building stock. There is also a developing culture associated with the conversion and renovation of existing buildings. Commercial owners now realize that their buildings are their major investment. Building surveyors are generally employed in private practice, local government and in many of the former nationalized industries.

General practice surveyors are employed in four main areas of work; agency, valuations, management and investment. Their knowledge and understanding of the local property market, land and property values are the particular attributes of this profession. Valuation is one of the main skill bases, being vital to investment work. These may be required for a variety of purposes such as sale, lease, insurance, investment or loans. General practice surveyors are employed in private practice and in government departments, and some are directly employed by property companies. The GP surveyor may be involved at the outset of a new project and is sometimes the client's first point of contact on a proposed development. They also advise the financial institutions on investment in order to yield the best result for their shareholders or members.

The Royal Institution of Chartered Surveyors

The Institution was formed in 1868 and incorporated by Royal Charter in 1881. It originated from the Surveyors Institute and has grown as a result of a number of mergers with other institutes, most notably the Land Agents (1970) and the Institute of Quantity Surveyors (1982). In 1995 its membership totalled 93 263 (see Table 10.2). The RICS is administered in seven divisions which represent the varying interests of different chartered surveyors. The General Practice (42%) and Quantity Surveying (37%) Divisions account for over three quarters of the membership. Chartered surveyors are employed in private practice (47%), government departments (21%) and with contractors or developers (16%). A larger proportion of quantity surveyors (16%) are employed with contractors than other types of surveyors. Eleven per cent are either non-practising or retired, and 10.5% of the membership are employed overseas, of whom the land surveyors (45%) and the quantity surveyors (16%) make up the largest membership. The RICS remains the largest of the all the professional bodies in the construction

Table 10.2   Divisional membership of the Royal Institution of Chartered Surveyors (1996)

	Fellows and Professional Associates	Students and Probationers	Totals
General practice	32 015	6 884	38 899
Quantity surveying	25 109	8 670	33 779
Building surveying	6 634	3 797	10 431
Planning & development	1 601	371	1 972
Rural practice	4 546	851	5 397
Land surveyors	1 147	723	1 870
Minerals	629	286	915
Totals	71 681	21 582	93 263

industry. It is also growing at a faster rate at the present time. During the 1990s, for example, its overall membership increased by 6433 compared to the growth of the Institution of Civil Engineers, the second largest by 3630 members.

Engineers

There is a wide range of different types of engineers employed both within and outside the construction industry. In civil engineering the design of the project is undertaken by the civil engineer. The powers of the civil engineer under the ICE Form of Contract, whilst they are comparable to those of the architect under the JCT80, are more wide ranging. In addition, the engineer's counterpart working for the contractor is also likely to be a civil engineer. The work of the civil engineer is diverse, often attempting to harness the powers of nature with projects typically costing several millions of pounds. The demarcation between building and civil engineering is ill-defined and there is overlap with some projects being a part of each. In order to emphasize their size and complexity typical projects include the Humber Bridge, Thames Barrier, Docklands Light Railway, Sizewell Nuclear Power Station, Electrification of East Coast Mainline and the Channel Tunnel. The latter was a joint venture with French civil engineers and contractors.

Structural engineers are really a branch of civil engineering concerned with the analysis of structural capability. They ensure that a building or other structure is able to withstand the different loads and pressures which might be applied. These include the dead loads such as the weight and forces of the structure itself and applied loads such as people, equipment, machinery (and the vibrations which they might cause), environmental factors such as wind pressures, the pressures caused through thermal expansion and contraction and even forces caused by earth tremors. Structural engineering is defined as the science and art of designing and making with economy and elegance, buildings, bridges, frameworks and other structures, so that they can safely resist the forces to which they might be subjected.

Structural engineers work closely with architects in the design of buildings, offering advice on the most appropriate types of foundations to be designed and the frameworks and supports which can safely carry the loads to which the building will be subjected. In addition, the structural engineer will undertake similar tasks with civil engineering structures. They also examine existing buildings, to comment upon their stability and ability to be used in circumstances for which they were not originally designed. As the shapes of structures in the modern world change, so too does the need for structural engineers to assess the different loads and behaviour of structures. Increasing use is made of computers and structural design packages which are able to model the structural deformations which will result under varying load conditions. Structural safety is a paramount consideration of the structural engineer's work. Structural engineers are typically employed in private practice but may pursue careers with contractors and in public service. A typical structural engineering consultancy firm employs between 25 and 50 staff.

The role and work of the building services engineer has changed from their obsolete title of heating and ventilating engineer. The change in title was not merely pragmatic but represented an ever-widening involvement in different types of buildings and in a far wider range of services for modern day buildings. The importance of providing comfort conditions in buildings represents a paradigm shift in concept and one which must be provided to achieve user satisfaction. Building services engineers claim that they are the people who bring a building to life. Services in buildings today include heating, lighting, plumbing, energy supply, telephones, computer systems, fire and security protection, etc. Without these services the building is a shell providing little more than the traditional function of shelter. Two trends have increased the importance of the building services engineers. Firstly, the increased knowledge and capability of designers coupled with improvements in technology. For example, the increasing importance in micro-chip technology has made such control systems possible. Modern buildings now use so much technology that they can be described as intelligent. As this technology becomes more commonplace its value in work percentage terms increases. In buildings 30% of the costs can often be allocated to services and in extreme examples this can rise to above 50%. Secondly, there is an increased emphasis on environmental issues. A function of the services engineer, in conjunction with the other members of the design team, is to design and construct energy efficient buildings.

Building services engineers are employed at the different phases during the life of a building. Initially, this involves working with the client and architect in establishing the brief and offering advice on how this might be translated into a services design. During construction the engineer inspects the systems to ensure that they have been installed in accordance with the specification and tested before the work is handed over to the client. The services engineer may be employed by the client to monitor and maintain the different systems in use and to suggest improvements and efficiencies as the technology evolves. Building services engineers are employed primarily in private practice, sometimes in group practices with other design professionals. They are also employed in central and local government and with commercial or construction companies.

The Institution of Civil Engineers

The term civil engineer appeared for the first time in the minutes of the Society of Civil Engineers which was founded in 1771. It marked the recognition of a new profession in Britain as distinct from the much older profession of military engineer. The members of the Society of Civil Engineers were developing the technology of the industrial revolution. There was no formal education even by the nineteenth century. In France, by contrast, the government had established the Grandes Ecoles to train engineers for the civil service. In order to remedy this situation in Britain, a group of aspiring engineers founded the Institution of Civil Engineers (ICE) in 1818. The most famous engineer of the day, Thomas Telford, became its first president and in 1928 it was formally recognized by the granting of a Royal Charter. This charter contains the often quoted definition of civil engineering as being 'the art of directing the great sources of power in nature for use and convenience of mankind'. The engineering committee of the Institution provides a forum in which engineers in different specialisms can exchange views and information. The designated areas include; ground engineering, water engineering, structural engineering, building technology, maritime engineering, transport engineering and energy.

In 1995 there were 79 606 members in the ICE of whom 49 726 were corporate members. The Institution of Municipal Engineers merged with the ICE in 1979. The engineering profession is very diverse with a variety of engineering institutions coming under the umbrella of the Engineering Council. Altogether there are about 300 000 members. In addition the Association of Consulting Engineers, acting as a voice for those in practice, has a membership which includes gas, chemical, electrical and mechanical as well as civil and structural engineers. Civil engineers work in the four main areas of contracting, consultancy, government and nationalized and denationalized industries. Consultancy accounts for the biggest proportion of members in employment, followed closely by those who work for contractors.

The Institution of Structural Engineers

The Institution of Structural Engineers (IStructE) began its life as the Concrete Institute in 1908, was renamed in 1922 and was incorporated by Royal Charter in 1934. Its aims include promoting the science and art of structural engineering in all its forms and furthering the education, training and competencies of its members. The science of structural engineering is the technical justification in terms of strength, safety, durability and serviceability of buildings and other structures. In 1995, the membership of the Institution stood at about 21 520 (12 008 corporate members and 9512 at some stage of training). The majority of structural engineers are employed in private practice.

The Chartered Institution of Building Services Engineers

The grant of a Royal Charter in 1976 enabled the Institution of Heating and Ventilating engineers, which was founded in 1897 to amalgamate with the Illuminating Engineering Society of 1909 to create an institution embracing the whole sphere of building services engineering. The objectives of the CIBSE are set out in their charter as the promotion of the art, science and practice of building services engineering for the benefit of all; and the advancement of education and research in this area of work. In 1995 there were 7495 qualified members in the CIBSE and a total membership of 15 363.

Builders

Builders are engaged in the administrative, commercial, educational, managerial, scientific and technical aspects of building. Seventy per cent of the CIOB members work in the private sector with contractors and subcontractors. The remainder are employed as consultants or work in the public sector. Almost half of those who work in the private sector are employed by contractors, and about 20% are employed in consultancy. The two main areas in which members offer services as consultants are project management (34%) and surveying and estimating (30%).

The Chartered Institute of Building

The CIOB was formed in 1834, incorporated under the Companies Acts in 1884 and granted a Royal charter in 1980. It originally started as the builders' society, which was a small exclusive club which had a wide influence and was responsible for helping to produce the early forms of contract. It is registered as a professional institution with, in 1995, 12 145 qualified members amongst a total membership of 32 370. Of the chartered bodies the CIOB has the largest proportion of non-qualified members. The objectives of the Institute are the promotion, for public benefit, of the science and practice of building, the advancement of education and science including research and the establishment and maintenance of appropriate standards of competence and conduct for those engaged in building. The Institute encourages the professional manager and technologist to work together with their technician counterparts in order to achieve a ladder of opportunity as a main objective of the training and examination structure of the institute.

Planners

Town planners are employed in the planning departments of local authorities throughout Britain and by many other statutory undertakings. Planners recognize that land is a limited and scarce resource and that the increasing demands placed upon it require an adequate system of allocation of use. Planners need to forecast changes in use but the pace of change in the twentieth century makes it difficult to plan for all eventualities. The effects of development are wide ranging. Consideration needs to be given to the protection of the countryside, the landscape, archaeological features, historic buildings, mineral reserves and water gathering grounds. The planner needs to consider aspects of whether or not a proposed development will cause pollution, increase road traffic, what effects it will have on neighbours and whether it will create even greater demands for further development.

The Royal Town Planning Institute

The Royal Town Planning Institute was founded in 1914. It is the recognized technical qualification for town planning appointments in public and private offices, both in Britain and overseas. In 1995 there was a total membership of 17 726 of whom 13 355 were corporate members of the institute.

Non-chartered bodies

The qualifications for entry into membership of a chartered institution is a university degree together with approved practical training. For those without degrees there is a number of non-chartered professional bodies with memberships ranging from as few as 1000 to nearly 10 000. There are over 20 different non-chartered bodies. Some of the larger ones are the Institute of Surveyors, Valuers and Auctioneers (ISVA 7500), the British Institute of Architectural Technicians (BIAT 5500) and the Association of Building Engineers

(ABE 6500). In recent years new associations have been formed to represent specialist interests such as the Institution of Civil Engineering Surveyors (ICES) or the Institute of Building Control Officers (IBCO). Some have been developed in association with institutions from other countries such as the Association of Cost Engineers which has strong links in the USA. Others are as a result of the changing needs and emphasis of the construction industry, with professional bodies such as the Association of Project Managers. The total membership of the non-chartered bodies is about 80 000.

Professional body membership

A proportion of members in the construction professions hold membership of more than one institution. It may be desirable, for example, to be a member of both the ICE and IStructE. Several members of the professions are also represented in other professional bodies which are more broadly based than construction alone. For example, a large proportion of the Chartered Institute of Arbitrators (CIArb) are members of the construction professions, whereas in the British Institute of Management (BIM) membership includes only a small proportion of those connected with the construction industry. In a relatively few cases a person may have originally qualified in a particular professional discipline but then developed an interest in a different area and taken up membership of a second institution whilst at the same time not relinquishing the former membership.

There is a proportion of individuals who are employed in practice in a professional capacity who hold no professional qualifications. Many of these provide a valuable service within their area of work. A small number having obtained a relevant degree, choose not to take up professional body membership believing that their degree is a sufficient qualification by which to practice. However, the aim of most students on built environment courses is to become members of their professional body. It is difficult to estimate the numbers involved in practice who do not have professional qualifications, but who would nevertheless still be described as a civil engineer, quantity surveyor, etc. (excluding those in a technician capacity). There are, allowing for retired members and those who no longer practice (those who have left the profession), about 400 000 (Table 10.4) who would describe themselves with a professional label of one sort or another. Of these over 25% are following a course of education prior to full professional membership. Total membership of the chartered professional bodies increased from 218 724 in 1982 (Table 10.3, Figure 10.1) to 292 846 in 1995, an increase of 34%. Some of this increase has been achieved

Table 10.3 Growth in chartered professional body membership [4]

		1982	1987	1988	1989	1990	1991	1992	1993	1994	1995	1995 % share
RICS	M	43 896	57 789	59 271	61 063	62 738	64 695	66 969	68 607	70 102	71 681	
	T	58 540	78 984	80 948	82 166	84 507	88 821	90 163	90 163	91 867	93 256	31.85
ICE	M	40 934	48 207	48 643	49 483	50 062	49 185	49 166	49 119	49 608	49 726	
	T	64 891	71 248	71 233	69 286	74 160	77 378	78 404	78 283	79 756	79 606	27.19
CIOB	M	8 894	9 250	9 458	9 642	9 901	10 719	11 065	11 436	11 863	12 145	
	T	27 517	27 538	27 585	28 382	30 009	32 270	32 529	32 575	32 372	32 370	11.05
RIBA	M	27 000	27 500	28 500	28 500	28 500	27 940	27 456	27 592	27 607	27 227	
	T	28 500	29 200	30 200	30 200	30 200	31 200	31 200	31 300	31 500	31 700	10.82
IStructE	M	10 771	11 211	11 316	11 531	11 771	11 703	11 927	12 201	12 442	12 513	
	T	15 015	17 930	19 111	19 333	20 465	20 692	21 566	22 577	23 010	22 825	7.79
RTPI	M	8 170	10 394	10 801	11 311	12 004	12 071	12 666	13 032	13 331	13 355	
	T	12 811	14 204	14 821	15 152	15 956	15 956	16 683	17 117	17 374	17 726	6.05
CIBSE	M	7 400	8 472	8 546	8 621	8 746	7 314	7 312	7 314	7 354	7 495	
	T	11 000	13 046	13 268	13 821	14 728	15 259	15 474	15 394	15 359	15 363	5.25
Totals	T	218 724	252 150	257 166	258 340	270 025	281 576	286 019	287 409	291 238	292 846	100.0

M = Corporate membership; T = total membership

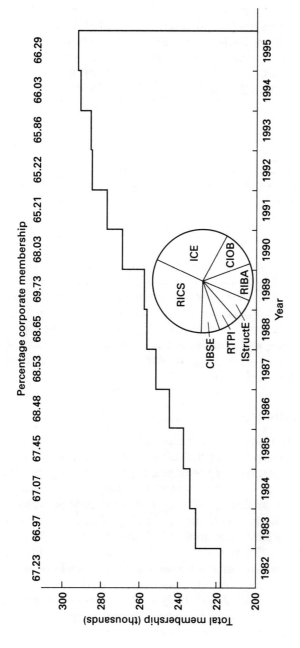

Figure 10.1 *Growth in chartered professional membership*

Table 10.4   Professionals in the construction industry

Chartered bodies	280 000
less dual membership	–10 000
Non-chartered bodies	80 000
less other membership	–40 000
Non members	100 000
Retired members	–15 000
Total	395 000

through the mergers of different societies. During the same period of time both the corporate and other classes of membership increased by a similar percentage.

Female membership

Women members are under-represented amongst the construction professions. Whilst most of the professional bodies have taken positive steps to increase the number of female members, the increases have been modest. In architecture, planning and estate management the numbers of female members are higher than in the other disciplines. In building and building services engineering it is small and, although recruiters are seeking to address this imbalance, they are achieving only limited success.

Europe and the USA

There are wide cultural considerations to be taken into account in any comparison between the construction industry professions in Britain with those in other parts of the world, notably, Europe, the USA and Japan. Historically, things have developed differently. In much of the rest of the world, architects and engineers dominate the construction industry. The ratio between building and civil engineering works in other countries is similar, at about 80:20. Construction GDP, however, varies from about 13% in Germany to 6% in Britain. The various professional disciplines in Britain are not mirrored elsewhere, other than in commonwealth and ex-commonwealth countries. The role of the professional bodies also varies. In Britain a professional qualification is one by which to practice. In Europe a professional body is more of an exclusive club, to which relatively few of those engaged in practice are members. In the USA there is the emerging discipline of construction management alongside those of architect and engineer. In Britain the architect is the only registered profession and this is also the case on mainland Europe, but its practice in Europe is more controlled. An architect, for example, must be employed to sign the plans in Europe, otherwise the building cannot be built. There is a growing body of opinion within Britain's construction industry that members of all professions should be registered [5] as precursory to practice. Interestingly, in some countries of the ex-commonwealth, such as Malaysia, registration is required for all of the main professions working in the construction industry.

Table 10.5   Architects and engineers employed in the construction industry [6]

| | | Per 1m population | |
	Architects	Engineers*	Total
United Kingdom	531	1815	2346
France	404	404	808
Germany	990	1639	2629
USA	250	1427	1677

* includes surveyors

Professional liability

The subject of professional liability is something that only a few years ago was a subject of little interest [7]. They were the days when the relationship between client and consultant was characterized by trust and confidence. From the 1970s onwards much occurred which eroded that trust, replacing it with conflict, uncertainty, risk and the possibility of litigation arising at some point. Also several other events have taken place to accelerate these issues, such as the NEDO building users insurance against latent defects report (1988), the DTI/DoE review of professional liability (1987), the harmonization initiative launched by the Directorate General III of the European Commission, and action by the Construction Industry Council in response to these initiatives. The liability for professional negligence has been well reported. Architects and engineers have already been required to make substantial payments for design faults or inadequate supervision, and surveyors have in the past been sued for negligent building surveys.

Construction Industry Council

The Construction Industry Council (CIC) is a body that has been established to represent the construction industry professions [8]. Membership includes the seven major chartered bodies (CIBSE, CIOB, ICE, IStructE, RIBA, RICS and the RTPI), business organizations which represent some of these professions and a number of the non-chartered bodies. Their objectives are as follows:

1   Membership of CIC by all professional bodies.
2   Generate effective relationships with employer organizations, trade associations and others.
3   Oversee standards development and co-ordinate NVQs relevant to members.
4   Promote the integration of compatible areas of professional education and training.
5   Broaden the income base of CIC.
6   Raise the profile, and increase understanding of the construction industry.
7   Promote continuing professional development.
8   Promote more extensive research and development.
9   Encourage improved industry practices and procedures and reduce duplication.
10 Represent members interests to government.
11 Co-ordinate and develop members interests in Europe.
12 Collect and disseminate information.

Consultancy

Consultancy work carried out by architects, engineers and surveyors can be analysed in several different ways. There are the firms who specialize in a single professional discipline, or even on a particular type of building or engineering project. Other firms adopt an integrated approach by employing specialists from a range of different professional backgrounds and also individuals who have no background in the construction industry. All of the larger practices include a range of the different specialisms rather than restricting themselves to a single professional background. Table 10.6 lists the largest of these firms based upon the numbers of staff that they each employ. W S Atkins is one of Britain's top consultants, with nearly double the chartered staff of its nearest rival. The Epsom-based firm completed the latest in a string of acquisitions with the Kenilworth based multi-disciplinary consultant DGI.

Predictably, engineering-led consultants dominate the top 100 table (published by *Building* magazine). But there is evidence that they are all diversifying. For example, nearly one-third of the staff at Pell Frischmann are not engineers. The league table also confirms the rise of the conglomerates, such as Chesterton International. This is a general practice surveyor that acquired the large quantity surveying practice of Cyril Sweett in December 1995. Others on the list include Rust, a subsidiary of the 17 000 strong USA-owned environmental engineer Capita Property Services. This firm also bought the quantity surveying practice of Beard Dove in October 1995.

The largest quantity surveying practice continues to be Davis, Langdon & Everest with 1515 staff and a world-wide staff of 719. Reflecting the moribund nature of the construction industry few of the top quantity surveying practices either opened or closed offices in Great Britain in 1994–95. Building Design Partnership, originally a architectural practice but now multidisciplinary has a world-wide staff of 780. It employs 157 chartered architects, a relatively few more than W S Atkins at 119.

During the recession years of the early 1990s there has been a significant amount of corporate activity among professional firms. Talks,

Table 10.6   Consultants by staff size [9]

	Total Worldwide 1995	Chartered Staff (UK)			
		Engs	Archits	Survey.	Project Managers
DTZ Debenham Thorpe	7 000	0	0	1 000	0
W S Atkins	4 840	1 506	149	272	85
Mott MacDonald Group	4 436	1 023	6	12	56
Ove Arup Partnership	4 156	634	12	15	40
Jones Lang Wootton	3 800	14	3	311	0
Acer Consultants	3 206	774	5	12	0
Pell Frischmann Group	2 753	658	61	111	94
Maunsell Associates	2 512	290	6	15	5
Rust	2 380	324	20	23	52
Sir William Halcrow & Partners	2 234	1 001	8	1	260
Tarmac Professional Services	2 125	325	90	75	110

mergers, acquisitions and joint working arrangements have frequently figured in the news pages of the trade and professional journals. In other cases firms have subdivided, where some of the partners have agreed on separate methods of working.

Two issues have dominated these firms. The percentage of firms forced to win 70% or more of their workloads on a competitive fee basis has stabilized at about 40%, with clients recognizing that they must make their appointments on the basis of quality of service as well as price. Consultants have also been busy in adding value to their firms in the eyes of their clients. Over 50% of consultancy firms are now certified to ISO 9000 (formerly BS 5750) with others in the process of acquiring this important kitemark.

In mainland Europe there are estimated to be in excess of 1500 offices of British consultants, the largest numbers of which are in Germany and France. Some of the consultants have a presence in many of the European capital cities. There is an anticipated growth in activity in the newly formed Eastern European countries and a major need for reconstruction in the former states of Yugoslavia after the ending of the civil war. The larger practices also have offices and undertake consultancy work in countries on every continent around the world.

The future of the built environment professions

The built environment professional bodies have grown steadily both numerically in membership and number throughout this century. The number of different professional bodies have continued to increase in spite of the amalgamation and mergers which have taken place. Some of the professional bodies such as the CIOB and the CIArb received Royal Charters during the 1980s. Others continue to press the Privy Council for this recognition. It can be argued that there is a proliferation of professional bodies in Britain. Often a member of a single professional discipline has the option of becoming a member of several different professional bodies. Even with registration, as in the case of architects, this only limits the use of the title. Whether this professional structure is beneficial to the client, the industry, the practitioners or the country at large is open for discussion [10]. The future of the professions in Britain is influenced by the following:

1   The effects of the Single European Market in 1992, since the structure of mainland Europe is different to that in Britain.
2   The diversification and blurring of professional boundaries often including other than built environment professions such as those involved with the law and finance.

3 Their role as learned societies.
4 The education structure of courses in the built environment.
5 The pressure groups both within and outside the construction industry.
6 The desire in some quarters for the formation of a single construction institute, to unify all professionals in the construction industry.

References

[1] Construction Industry Training Board. *The Construction Industry Handbook*. Hobsons Publishing, (1991)
[2] Routh, G. *Occupational pay in Great Britain. 1906–79*
[3] Ivanhoe guides to: Chartered builders, Chartered surveyors, Chartered architects and the Engineering professions. The Ivanhoe Press
[4] Professional body membership records
[5] NEDO. *Registration in the Construction Industry*. 1992
[6] Department of the Environment. *Professional Education for Construction: Overseas comparisons*, 1989
[7] Massey, A. and Manson, K. 'Professional liability and indemnity insurance' in *Chartered Builder*, November 1990
[8] Construction Industry Council. *Corporate Plan*, 1992
[9] New Builder Professional Services File, 1995
[10] IPRA. *Future skill needs of the construction industries*. IPRA, 1991

11

Management and supervision

Yet still there are signs in all sorts of ways that the industry is short
of management ability and it would be difficult to find any knowl-
edgeable person in the industry to assert that the standard of management overall is satisfactory.

P.A. Hillebrandt, 1988.

Parties involved in construction management

As shown in Chapter 6, the contracting firm varies in size from a few
employees to businesses with several thousand employees. Most are
owned and managed by a sole proprietor or a partnership; others are
companies, both private and public. For the bigger organizations, the
chain of control will involve many layers of management between the
operative and the boardroom, for example, ganger, trade foreman,
general foreman, agent, contracts manager, general manager and
director. This compares with the small firm in which the proprietor is
actually on the site or controlling it through a foreman. In the large
firms the managerial functions are departmentalized. On average,
administrative, professional, technical and clerical employees represent some 20% of the total labour force in contracting firms, with the
larger firms employing the greater numbers. Table 11.1 and Figure
11.1 show the distribution of the administrative, professional, technical and clerical staff. It can be seen that the number of managerial
staff has increased by nearly 13% during the 1980s, although there
was little change in the total number of the professional workforce.
Until the late 1980s there has been an overall decline of 26% since
1990.

Table 11.1   Administrative, professional, technical and clerical staff in contracting firms [1]

	1980	1981	1982	1983	1984	1985	1986	1987	1988	1989	1990	1991	1992	1993	1994
						(,000s)									
Managerial staff	61.3	60.1	58.8	60.0	59.3	57.0	56.5	54.5	66.0	69.2					
Architects, surveyors and engineers	18.3	16.7	16.8	17.1	17.3	17.4	17.1	18.0	17.9	19.0					
Technical staff	27.6	26.4	25.2	24.7	24.7	24.9	24.6	27.6	31.4	32.3		Details			
Draughtsmen and tracers	3.2	3.0	2.8	2.7	2.8	2.6	2.5	2.5	2.8	2.9		discontinued			
Foremen	38.1	34.7	34.0	33.6	33.8	32.1	32.2	31.0	29.6	30.1					
Clerical & sales staff	90.7	85.3	82.2	82.5	82.6	78.7	78.9	77.5	79.5	82.4					
Totals	239.2	226.2	219.8	220.6	220.5	212.7	211.8	211.1	227.2	235.9	266.0	251.0	221.0	204.0	198.0

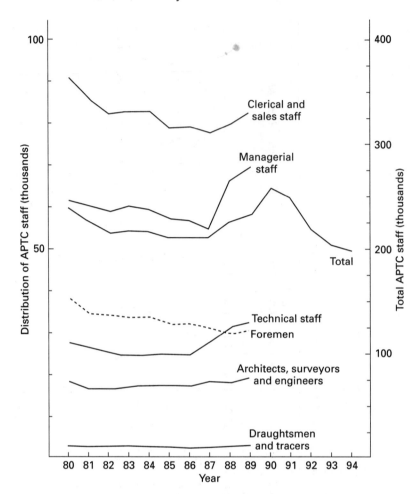

Figure 11.1   *Administrative, professional, technical and clerical staff*

During the 1970s there was said to be considerable improvement in the management skills of senior management with better training and more qualifications available in the management field [2]. Nonetheless, it was judged that the major constraint of firms is the lack of good project and site management. Many of the smaller contracting firms are specialist subcontractors. There are management problems in these small firms and considerable scope for training more of these managers.

A survey carried out into the promoters' opinions of the construction industry [3] indicated that poor management was a cause of complaint. Contractors were often felt to have insufficient control over subcontractors. An allied and frequently mentioned constraint on the effectiveness of the construction industry was the lack of co-ordination. It was felt that the industry was badly organized and never really did a satisfactory job. For example, materials and fittings were not co-ordinated for delivery. One promoter, a large firm, considered that the construction industry was very bad at programming and had little experience or expertise in process-intensive work. In that firm very precise programming was required to co-ordinate plant installation and only a few national contractors offered adequate process experience. The importance of adequate supervision, whatever the approach to construction, was one of the main points stressed, particularly by larger promoters.

Some promoters expressed dissatisfaction with the traditional role of architect as project leader. As managers, it was judged that they had shortcomings. One large firm complained of the uncertain responsibilities of architects and quantity surveyors. Others complained of the architects' remoteness from the building process and the site, in addition to their lack of industrial experience. Little seems to have changed from an earlier report [4] in which the following passage is quoted:

> there is all too often a lack of confidence between architect and builder amounting at its worst to distrust and mutual recrimination. Even at their best relations are affected by an aloofness which cannot make for efficiency, and the building owner suffers. In no other industry is the responsibility for design so far removed from the responsibility for production.

The organizational structure

The parties in the construction contract which carry the financial risk are the promoter and the contractor. As the design function is often separated from the construction function, the traditional mode of operation in construction has been to place the designer between the promoter and the contractor. The designer is not involved in the contract which exists between the promoter and the contractor but undertakes some of the planning and control functions. If the promoter is the contractor, for example, the undertaking of a speculative building project which the contractor is intending to sell or rent, the process may be relatively straightforward. The designers are the construction professionals and may, in a large project, be an architect, quantity surveyor and engineer. In addition to the main contractor, nominated subcontractors may be used. Most likely, there will be other subcontractors operating. The promoter makes separate contracts with the quantity surveyor, architect, engineer and the main contractor. The main contractor, which is the key element of the production team, makes separate contracts with the subcontractors. The subcontractors may make further contracts with others. Sometimes part of the works may be subcontracted several times. The separation of the design and production processes can lead to design which cannot be easily constructed; if there is no feedback to the design team, designs with these details can be repeated for other projects. In comparison with the USA, Europe and Japan, the British construction industry is noted for the relatively long construction periods and relatively high costs [5].

Better control may be exercised over the project by placing a project manager in overall control. This centralizes policy and decision-making which assists in co-ordination between design and production. The project manager can be an individual or a company depending on the type of contract. A further development is the use of a management contractor which is an individual or firm engaged by the client to manage the construction process. A large project often uses a management contractor who will be paid a fixed fee plus reimbursible

Table 11.2   The effect of method of organization on the distribution of site times and total times for the case studies investigated [6]

Method of organization	% projects with site times			% projects with total times		
	Fast	Average	Slow	Fast	Average	Slow
Traditional	17	37	46	25	29	46
Own management	20	40	40	40	20	40
Design and build	47	11	42	47	21	32
Separate management function	50	38	12	63	12	25

Fast projects are those completed at least 10% faster than the sample average.
Average projects are those completed within 10% of the sample average.
Slow projects are those completed at least 10% slower than the sample average.

costs. The management contractor will enter into contracts with the contractors and works with the promoter, the project manager and the design team from inception to completion of the project. Finally, the all-in, or turn-key, contract constitutes a departure from the foregoing procedures. In the all-in contract, the promoter states his requirements in broad and general terms and invites contractors to submit proposals for the project. This includes design and construction. Payment can be by several systems, sometimes a fixed price contract is undertaken. The procedure has been in use for many years in engineering construction, for example, steelworks, chemical and oil industries.

A detailed investigation was made of a sample of 56 case study projects [6] which were classified by the method of contract organization. The method used was related to (i) the relative speed of construction on site which was defined as the percentage difference from the average time obtained from the sample for projects of a similar size, and (ii) actual total times from inception to completion and, as before, the percentage difference of the total time from the average time. The results are shown in Table 11.2 for the different methods of contract organization. A higher proportion of projects in which non-traditional methods of contract organization were used had fast times.

Some individuals in the management chain

In engineering work it is usual to appoint a resident engineer who is the representative of the engineer. The resident engineer's duties are not restricted to inspection and supervision. He has complete control and authority to make decisions and to act in all ways on behalf of the engineer. This includes issuing instructions and drawings to the contractor's agent. This degree of control is necessary because much engineering work is carried on in remote places where it is not possible always to refer day-to-day problems to an office in Britain in the same way that building site problems can be referred to an architect. In engineering work the problems that arise, although less in number than those found on an ordinary building site, are usually more serious and complex and call for decisions quickly with no opportunity for reference to others. Day-to-day supervision also calls for professional qualifications and ability to make calculations and so forth which would be beyond the capacity of a clerk-of-works. In major works the resident engineer will probably have under his control a considerable supervisory staff.

The contractor's representative on site is the engineer in charge, generally described as the agent. The agent is usually an experienced engineer who is given wide discretionary powers. The agent's main duties are to ensure that the works are carried out economically and in accordance with the contract documents to the proper requirements of the engineer. The resident engineer and the agent for the most part work in close co-operation. Where the work involves installation of equipment which is being supplied by a concern not party to the main contract, an independent inspecting engineer is frequently appointed. He has obligations to the promoter, the contractor and the equipment supplier to ensure that no dispute subsequently arises over the quality and suitability of the equipment or mode of installation.

The counterpart of the resident engineer in a building contract is the clerk-of-works, i.e. the site representative of the designer. The clerk-of-works' duties are to act as inspector on behalf of the promoter under the direction of the architect. The clerk-of-works' directions to the contractor or his foreman are ineffective unless confirmed in writing by the architect. The extent of the powers and duties of the clerk-of-works and of the architect's responsibility for his shortcomings has been the subject of much litigation. Yet, in practice, an investigation [7] by the Building Research Establishment suggested that the role of the clerk-of-works is crucial to the construction of satisfactory buildings. The qualifications for clerks-of-works are that they must have had an extensive trade experience on construction sites. Apart from those employed in government departments as full-time employees, it is unlikely that they will have the benefit of courses in supervision and management. A comprehensive report [8] by the National Health Service emphasized the need for an improved educational programme for clerks-of-works in the Health Service. Some architects employ full-time clerks-of-works who are available to supervise any work for which the firm is responsible; in other cases the clerk-of-works may be a casual employee taken on for a particular job.

The general foreman is an important link between the site management and the foreman and gangers in direct charge of the labour. He directs labour, materials and plant to suit the construction operations and his competence may determine whether the work will involve the contractor in loss rather than profit. General foremen have, for the most part, served as apprentices, trade foremen and have obtained their position after long experience on construction sites. It is likely that only relatively few have had the benefit of a course in site organization and management. In 1950, it was estimated that the number of general foremen was rather more than 8000 [9]. The availability in 1973 was approximately 58 600 and the highest and lowest requirements in 1983 were estimated at 78 000 and 41 700 respectively [2]. These estimates were based on one general foreman employed for approximately 13 skilled operatives directly employed by contractors or public authorities. As subcontracting has increased the lowest level of site control for the contractor has been raised above that of the sub-contractors for which the general foreman now operates, reflecting the interests of the subcontractor rather than that of the contractor. Table 11.1 shows that, during the 1980s, the number of foremen has fallen by over 20%; no doubt, many of them turned to self-employment.

As projects have become larger and more complex, it has become the practice not to use the designers as managers of the contract but to place the overall control in the hands of a project manager. Management contracting is also used to manage a contract. In this case the design remains independent of the construction but the role of the main contractor is replaced in so far that the management is exercised by the management contractor. Despite being limited to the construction rather than having control of the design as well, the advantage is that the management contractor is involved in the planning of the works from an early stage. The construction work on the site can still be let on a competitive basis to the subcontractors.

Incentive schemes and motivation

Bonus payments have been used in the construction industry for many decades. They have been the cause of much conflict between operatives and management yet are believed to result in a more quickly completed job, thus offsetting the additional wage bill. Benefits are also thought to arise from the resulting improved site organization and management as operatives who are earning bonuses will not tolerate working practices and conditions that cause delay to their work.

The basis of a good incentive scheme is that it should give an operative of reasonable ability an opportunity to increase his income above the basic wage, say 30% more, in return for increased production. The formulation of such a scheme requires both technical and psychological skills. Whether or not the operatives benefit significantly from the bonus system is arguable. The construction trades unions believe that incentive schemes tend to depress the basic rate of payment and it is said that some would welcome a higher basic rate linked to the discontinuance of incentive schemes.

Despite the fact that there is a number of motivating factors for people in employment, construction management apply particular attention to pay and less attention to non-financial incentives where operatives are concerned. In the case of salaried staff there are often company cars, pension schemes and other benefits offered as incentives.

During the late 1960s and early 1970s, public attention was drawn to the avoidance of income tax payments by some self-employed labour-only construction operatives. Paying the self-employed in full was seen to be advantageous to both operatives and construction companies. The former were getting a high wage although they were not entitled to benefits which other employees received; the latter benefited from low overhead costs, the flexibility of labour and a ready-made incentive scheme. However, the British Government was deprived of some £300m per year of lost tax. No doubt, over the years, this has done substantial damage to the overall prosperity of the nation. With the widespread growth of self-employment in construction the Inland Revenue is more alert to tax avoidance. Nonetheless, the trades unions are concerned that self-employment seriously undermines the wages and conditions of the directly employed, more especially on training and safety standards although there is no evidence to indicate that the self-employed are more at risk.

Fast track construction

To reduce the overall completion time for a project, a management technique, recently called fast tracking, has been used. It involves a system which uses an integration of procedures and processes which, although more expensive to implement satisfactorily, can assure the promoter of an early occupancy. Some of the main features of the technique are as follows:

- The design and contracting functions are largely overlapping rather than using the traditional system. The design stage is not completed prior to tendering and construction. Instead, the contract is let on the basis of preliminary information, first-stage quantities or rates.
- Construction starts fairly soon after the design work is commenced and foundation work will be undertaken before much of the detailed design. Only later will the detailed design of all the parts of the construction, for example, the upper floors, become available. In order to operate in this way, the whole process is brought under the authority of a project manager or, alternatively, a management contractor is used. Either way, the contractor is involved in design and construction phases.

- The construction is planned to be undertaken using approaches akin to those used in manufacturing assembly lines. For example, the use of repetitive work sequences using standard off-site fabricated elements and eliminating, as far as possible, delay in the use of wet trades.
- Delivery of components and materials is based on the 'just in time' inventory control techniques. This avoids unnecessary handling and storage.
- High investment in site mechanization is made to ensure the fast and efficient transport of men and materials to the workplace.
- Design and erect subcontracting is encouraged.

High quality personnel are necessary to ensure that the system works effectively without generating delays. Work variations are kept to a minimum. It is reported [5] that the extra costs involved in fast tracking are some 7.5%; this is greatly offset by earlier completions and a promoter may be willing to pay for a construction system which has a higher capital cost but faster construction time. This is one reason why structural steelwork is favoured for multi-storey buildings rather than *in situ* reinforced concrete. The developer of the Broadgate Office scheme in the City of London (1984–86) claimed that fast tracking saved over 30% of the construction costs. In addition, rents were received a year earlier and substantial interest charges on work in progress were avoided [10]. Mahaffey [11] showed that, during a construction period of some 8 months for a multi-storey building in an Australian city, the associated charges including interest, rates, taxes, insurance and administration amount to some 8% of the cost of the land and new building.

Some of the characteristics of fast tracking are similar to those of the all-in contract, turn-key contract, and package deal undertaken by large civil engineering contractors. Fast tracking can operate satisfactorily only when a high level of managerial expertise is available. Often the management is not able to operate fast track procedures due to traditional procedures exercising constraint, lack of commitment by the client or insufficient managerial skills.

The role of new technology

The computer and its associated technologies have been shown to be effective tools for a wide range of operational and managerial uses. In particular, the use of computers in the design professions has shown considerable success in the representation and analysis of design solutions. Although information technology applications are capable of achieving high work levels and have been reported to offer time savings of up to 40%, they have failed to meet the expectations for increased productivity and product quality in the construction industry. The opportunities are substantial; for example, the following areas lend themselves to the development of information technology in construction applications:

- design and production techniques which incorporate design aids, robots, energy management, commissioning of buildings and education and training
- information systems which employ data bases, quantities, drawings and models, specifications, property data and electronic data interchange
- hardware and software which include interfaces, expert systems, standards, integration of applications and software techniques
- communications which apply to intelligent buildings, wide area networks, local area networks, integrated services, digital networks, optical fibres and wiring, radio technology and security

Despite the opportunities for the greater use of information technology in the British construction industry, a report [12] published in 1990 states that there had been little extra increase in the uptake of information technology for 2 years. Seventy per cent of the top 400 contractors who responded to the survey are still frustrated by the lack of appropriate hardware and software. There is also a lack of trained personnel to use the equipment and a lack of understanding about computers by the top management. However, the survey indicated that, in contrast to the contractors, an increasing number of consultants are using computers. Of the architects who responded to the survey, some 80% stated that they used computer-aided design equipment. Overall, in British practices the use of computer-aided design is estimated at approximately 20%. Only 4% use electronic mail.

Overall, construction spends less on information technology than almost any other industry. The financial services sector invests 5% of its annual turnover, manufacturing sets aside 2% but construction only 0.5%. Such statistics have contributed, in part, to the suggestion that the construction industry is backward [13]. Its technological growth has been retarded and it exists largely as a nineteenth century handicrafts technology.

Health and safety

There has been concern shown for many years over accidents and working conditions in construction work. A Public Enquiry was held in 1904 and the first Building Regulations were passed by Parliament in 1926. These regulations were revised and updated in 1931. The 1937 Factories Act empowered the Minister of Labour to make regulations for the building industry but it was not until 1948 that the Building (Safety, Health and Welfare) Regulations were brought into force. These regulations did not apply to civil engineering works. During the post-war years the construction industry expanded and the accident rate gave cause for concern. In March 1962, the Construction (General Provision) Regulations and the Construction (Lifting Operations) Regulations replaced parts of the earlier regulations. These were extended to civil engineering as well as building. Two further sets of regulations, the Construction (Health and Welfare) Regulations 1966 and the Construction (Working Places) Regulations 1966 were brought into operation. In 1974, a major piece of legislation, the Health and Safety at Work Act 1974 was introduced. Further regulations have since been added. The Health and Safety at Work Act covers all people at work with few exceptions whether employers, employees or the self-employed. It also protects the general public in cases where the work activities of others may affect its health and safety.

Despite the regulations, and the certainty that any injury resulting from their infringement will lead to criminal prosecution and a civil action for damages, the accident rates in the construction industry are high. At a recent conference where these matters were discussed, it was judged that the accident rates in the construction industry were unacceptable. Steps must be taken to reduce the rates to lower levels, at least compatible with those of other developed countries [14].

Table 11.3 shows the annual deaths, major non-fatal injuries and injuries resulting in an absence from work of over three days per 10 000 persons, i.e. employees and self-employed due to work in the construction industry. Were members of the public to be included, the fatalities would be increased by some 10%. For comparison, corresponding statistics are shown for the manufacturing industry. Fatality data are plotted in Figure 11.2. There were, on average, two deaths every week on construction sites during recent years. It is judged that 90% of these deaths could have been prevented. In 70% of cases, positive action by management could have saved lives [15].

The risk of a fatal injury in construction is about three times greater than in the manufacturing industry; the risk of a major injury is 1.3 times greater in construction. Of the fatal accidents, 74% occurred in the building sector of the construction industry, i.e. construction, demolition and maintenance of houses, commercial and industrial premises. Some 26% were associated with the civil engineering sector

Table 11.3 Deaths and injuries per 10 000 persons due to work in the construction and manufacturing industries [16]

	Year													
	80	81	82	83	84	85	86/87*	87/88	88/89	89/90	90/91	91/92	92/93	93/94
Construction industry:														
Deaths	0.79	0.79	0.88	1.05	0.88	0.93	0.83	0.90	0.82	0.85	0.68	0.58	0.60	0.62
Major injuries	na	na	10.7	12.1	12.8	11.8	21.3	21.2	21.8	22.7	19.9	19.5	18.0	18.2
Over three days	na	na	na	na	na	na	115	111	105	102	99.6	95.6	83.6	78.4
Manufacturing industry:														
Deaths	0.19	0.22	0.23	0.22	0.25	0.22	0.22	0.20	0.20	0.21	0.18	0.18	0.17	0.19
Major injuries	na	na	6.9	8.0	8.3	8.4	14.7	14.4	14.7	14.8	15.0	14.2	13.8	14.4
Over three days	na	na	na	na	na	na	106	103	110	119	123	125	120	118

* From 1986 data were collected for the year commencing 1st April. More recent figures are for accidents rather than injuries.

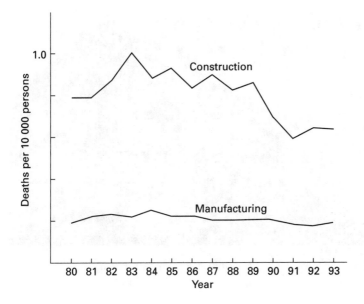

Figure 11.2   *Deaths per 10 000 persons*

(see Figure 11.3), i.e. construction and maintenance of roads, sewer pipelines, sea and harbour defences and large petrochemical, oil and gas installations [17]. The accident level for fatalities is reducing in construction and the major injuries have decreased in the past few years. Because of the nature of the construction industry, there are particular hazards for those who work in it (see Figure 11.4). The working environment on a construction site is subject to constant change and, as the working activities change, new risks are involved. The casual and short-term nature of employment compounds these risks. In contrast, most manufacturing is undertaken in a more controlled environment with specified activities undertaken by a relatively unchanging workforce. Were the nature of construction to change, with a greater bulk of component parts fabricated under more closely controlled conditions and brought to site for erection only, safety benefits could result. Increasing industrialization has benefits also for quality and productivity.

Further contributory factors to the higher incidence of accidents in construction could be the multiplicity of small firms and the widespread system of subcontractors.

Safety management may not be high profile, particularly if competition is keen and prices low. Under these circumstances, accident

Figure 11.3   *The tragic Milford Haven Bridge where four men died*

statistics involving other than major occurrences might not be reliable. There may be failure to comply with reporting procedures leading to underestimation of the total injuries incurred in the industry.

Table 11.4 shows the causes of fatal injuries in the construction

Figure 11.4   *Working conditions can be far from ideal*

Table 11.4 Causes of fatal injuries in the construction industry (1981–85) [15]

Falls (of people)	52%
Falling materials/objects	19%
Transport/mobile plant	18%
Electricity	5%
Asphyxiation/drowning	3%
Fire/explosions	2%
Miscellaneous	1%

industry for the years 1981–85 [15], and is an analysis of 681 deaths. The bulk of the fatalities are due to falls and being struck by falling objects. Not unexpectedly, these are also the main causes of the major injuries.

The Health and Safety at Work Act 1974 also requires that the employer identifies health hazards associated with construction and takes adequate measures to protect employees and others who could be affected. Hazards can be of a chemical nature and include such substances as paint solvents and other materials that are dangerous due to contact with the skin, inhalation or ingestion. Physical hazards, for example, cold and vibration, and biological hazards are also included as potentially injurious to health. The regulations apply to health hazards for which exposure can have an immediate or delayed effect.

The maintenance of health and safety on construction sites is sensitive to the competence of the management and the attitude of those exercising control over the construction activities. Commitment must be made by all parties contributing to design and construction to take appropriate measures to protect the health and safety of persons affected by the works. In reference [15] it was stated that:

Better management of sites through detailed pre-site planning with all who are to be involved in the job is needed to improve the general level of safety. This requires discussion with architects, engineers and other professional advisers, as well as main and subcontractors, safety representatives and safety professionals. Coordination of the work, with particular attention to high risk activities, can reduce the overall risks.

References

[1] Department of the Environment. *Housing and construction statistics.* HMSO, London

[2] *Construction into the early 1980s. The implications for manpower and materials of possible levels and patterns of demand.* HMSO, London, 1976

[3] NEDO. *Construction for industrial recovery. The role of building and civil engineering in promoting the British Manufacturing Industry.* HMSO, London, 1978

[4] Ministry of Public Building and Works. *Survey of problems before the construction industries.* HMSO, London, 1962

[5] Kwakye, A.A. Fast track construction. *Occasional Paper No. 46.* The Chartered Institute of Building. 1991

[6] National Economic Development Industry. *Faster building for industry.* HMSO, London, 1983

[7] Bentley, M.J.C. 'Quality control on building sites', *CP7/81.* Building Research Establishment, 1981

[8] Clark, T. 'Building clerks of work in the NHS: towards an educational policy', *CEU Working Paper.* York Institute of Advanced Architectural Studies. 1982

[9] Ministry of Works. *Working party report. Building.* HMSO, London, 1950

[10] *Financial Times,* October 1986

[11] Mahaffey, P.I. 'Design for construction'. *Proceedings of Concrete Symposium.* Cement and Concrete Association of Australia. Brisbane, August 1977

[12] *Building on IT for the 1990s.* Peat Marwick and McLintock. Computing Association, 1990

[13] Clarke, L. 'The production of the built environment: backward or peculiar?' in *The Production of the Built Environment,* 6. Bartlett School, University College, London, 1985

[14] 'Economic Construction Techniques'. *Conference Proceedings.* Institution of Civil Engineers. London, 1988

[15] Health and Safety Executive. *Blackspot Construction.* 1988

[16] Central Statistical Office. *Annual Abstract of Statistics*

[17] Davies, V.J. and Tomasin, K. *Construction Safety Handbook.* Thomas Telford, London, 1990

12

Planning, programming and progressing

Technological and administrative methods used during construction should be improved to achieve the greatest economic efficiency.

Pier Luigi Nervi, 1956

Construction planning

The construction process includes all activities involved in bringing to fruition a construction project from the conception of the project to its physical completion on the site. This involves three major phases including the conceptual; design and contract documentation; and construction. The time required for the construction process varies with the size of the job and can take only weeks for repair and maintenance work to over a decade for large projects.

Chapter 2 indicates that the time required for pre-construction site activities is less predictable than that for the site work. The pre-site phases also occupy a longer period of time. The overall duration of the construction process makes it difficult to suddenly increase or decrease the output of the industry. However, a measure of improvement in the speed of the process can be achieved by merging the latter two phases using fast track procedures (see Chapters 3 and 5). When completed, some public sector works such as roads, water and sewerage plants will be utilized over periods exceeding 25 or 30 years.

The planning of construction works is fraught with difficulties. Even a relatively small project needs careful planning in order to try to ensure the uninterruption of work and achieve a balance between progress, quality, safety and economy. Although planning will not solve the problems of inclement weather, delay in the delivery of materials and plant breakdowns, it will provide a flexible guide in order to compare the work planned against that actually achieved.

For each objective there is an established planning and control system. These are shown in Table 12.1.

The construction industry has many examples of cases in which

Table 12.1   Planning and control systems [1]

	Planning systems	*Control systems*
Progress	Programming	Progress control
Quality	Drawings	Quality control
	Specification	
Safety	Law on safety	Safety statistics
	Contract clauses	
	Site instructions	
Economy	Estimating	Cost control
	Budgeting	Budgeting control
	Cash flow forecasting	Cash flow control
	Profit forecasts	Financial control

planning was weak, i.e. deciding what is wanted and how it is to be achieved, or where control was ineffective, i.e. comparing what has been done with what ought to have been done. Examples of the former include the bridge disaster at Milford Haven and of the latter the Ronan Point collapse. However, neither of these cases are clear-cut. When disasters occur there is usually a multiplicity of interrelated factors which should have been detected by suitable project control and management procedures. Pugsley [2] identified the 'engineering climate' which relates to the atmosphere surrounding the conception, design and use of a project, which can cloud personal objectivity and lead to error.

Planning Techniques

A programme or schedule is developed by breaking down the work involved in a construction project into a series of operations which are then shown in an ordered stage-by-stage representation. Without a programme of work which specifies the time and resources allocation in order to undertake each stage of the project, the execution of the contract will be haphazard and disordered. The several methods used in programming may be broadly classified as follows: cumulative progress charts (S-curves); bar charts; networks. The S-curve is the simplest chart. The bar chart is the best known of all planning techniques. In its simplest form the sequential relationships between activities are not completely prescribed. However, they can be linked to show the relationship between an activity and the preceding and succeeding activities. Thus, dependency between the activities highlights the effect of delays. The resources required can also be calculated. If the bar chart is produced manually, and not by a computer system, it is limited in scope and not easy to update.

The linkage between preceding and succeeding activities combined with a set of arrows to represent the bars of the bar chart gave rise to a simple network diagram. This formed the basis of a network analysis, i.e. the critical path method, which identified the longest irreducible sequence of events and also lent itself to manipulation by holding the data in computer files. This results in a powerful planning technique which can be quickly updated. It also defines quickly those parts of the programme which could benefit from the use of increased resources and thereby benefit the project. As networks are rarely the best method for communication, the output of the analysis is often presented as a bar chart. Network analysis has been used for many large and complex projects. It has been claimed that time savings in excess of 40% have been achieved [3] by its use.

As planning has become a more scientific and specialist task, it has generated a recognizable cost in its own right. There is little in the way of published data on planning costs but it is clear that the more sophisticated planning techniques are the most expensive in terms of first cost. For large, complex projects there is probably no alternative but to use the most advanced planning techniques.

Table 12.2 summarizes the approximate costs of programming. On some jobs, or parts of jobs, the penalties of overrunning the contract

Table 12.2   A first stage assessment of the costs of programming [1]

	Total cost of project (£m)				
Method of programming used	0.01	0.1	1	10	100
S-curves	0.3%	0.2%	0.1%	0.05%	0.05%
Simple bar charts	0.5%	0.4%	0.2%	0.1%	0.1%
Complex bar charts	0.7%	0.6%	0.3%	0.2%	0.2%
Network analysis	1.0%	0.8%	0.5%	0.4%	0.4%

completion date are so severe that the information below will not apply. Under such circumstances significant sums may be spent on programming to yield the full potential benefit.

The features of a good programme are that it is flexible enough to allow modification to meet the contingencies that arise. Unexpected events can occur and the originally planned sequence of events is disturbed. Under these circumstances a well designed programme is of value as it demonstrates the effect that the contingency has on the events as originally planned and adjustments can be made accordingly.

Table 12.3 has been compiled from an investigation of construction planning for industrial buildings. It gives an indication of the reasons for the delays that occurred and the frequency of their occurrence.

Table 12.3   Delays in industrial buildings [4]

Cause of hold-up and delay	Case studies % projects
Subcontracting	49
Tenant/client variations	45
Ground problems – water, rock, etc.	37
Bad weather	27
Materials delivery	25
Sewer/drains obstruction or re-routeing	20
Information late	20
Poor site management and supervision	18
Steel strike	16
Statutory undertakers	14
Labour shortage	10
Design complexity	6

The influence of these delays on the total site time varied. For example, ground problems or bad weather occurred on sites which achieved both fast and slow site times suggesting that there was less overall effect. However, poor site management and supervision, late information and deliveries, design complexities, difficulties with statutory undertakers and labour shortages were usually associated with construction overruns of two months or more. These factors were judged to be the most damaging to the more effective organization of the construction process.

Resources

The time taken to complete an activity in the programme is dependent on the resources allocated to that activity. The approaches used in assessing the required resources can be based on completing the project in a given time or completing the project with specified limited resources. Once the level of resourcing has been finalized, the overall resource demands are smoothed if necessary by re-scheduling activities to ensure an acceptable overall demand for the project.

There is a measure of uncertainty in estimating the time for each activity particularly as delays in delivery of materials and adverse weather can delay work in progress. Probabilistic distributions have been used for the generation of the most likely times for activities; an approach that was used in early projects was the project evaluation and review technique (PERT).

Generally there should not be difficulty in ensuring that the materials, properly ordered in good time, are delivered to site in time to undertake the required activity. Some materials have long lead times and these may prove expensive for early delivery; however, an accommodation can sometimes be found by agreeing a substitute material. A less conscientious contractor may decide to ignore the difficulties and seek to obtain a claim at the end of the contract. The position relating to the flexibility of the workforce has changed in recent years. When contractors ran a number of sites in an area with their own workforce forming the bulk of the manpower, there were ample opportunities for operating flexibly. However, the arrival of a greater measure of subcontracting has meant that workforces are less flexible than a directly employed workforce. There is less flexibility in day-to-day site management and adaptibility has been replaced by a greater measure of pre-site planning.

The labour-intensive nature of contracting has brought about a state of affairs where the use of plant and machinery in the construction industry has traditionally been less than that used in manufacturing. Nonetheless, techniques do require the use of plant in which firms have directly invested or obtained by means of plant-hire. The equipment used is necessarily mobile and down-time or idle-time is minimized to ensure that the return on the investment is achieved. Nonetheless, the amount of time that plant is unused can be high; as much as 90%. This places a considerable capital burden on the contractor. The plant-hire industry ensures the more intensive use of plant and the hirer also supplies skilled operators. Plant-hire is responsible for supplying over half the needs of the construction industry.

Monitoring and control

With a programme of work and the resource requirements for each activity having been determined, it is possible to monitor the construction work as it progresses. In practice, updating will take place and control will be exercised. This will entail the re-scheduling of activities and the revision of resources. The emphasis on greater control, particularly cost control and site activities, has been one of the more important developments in the construction industry during the past decade or so. This was brought about by the decline in workload in the early 1970s which caused contracting firms to focus more closely on their profitability, thus paying greater attention to site management and control.

The planning model is often used to explore the overall development of the project before work on site is undertaken. This assists in investigating the influence of different construction techniques and the timing of the individual activities to optimize the use of resources.

Ensuring that the works are constructed to the specified level of quality is essential. This extends from the initial setting out of the project to the inspection, storage, handling and incorporation of the specified materials. All site operations should be governed by appropriate safety measures which start with a safe design and erection procedure for permanent and temporary works. Safety control is a matter of comparing site statistics with comparable statistics in addition to locating areas where it is necessary to take action. This is not at all easy to do. Construction is the largest single contributor to fatal accidents under the Factories Act. It was calculated in 1970 that a single reportable (three day) accident cost on average about £300, though one single major accident might cost £500 000 [1]. Comparable figures in 1992 might be £2500 and £2m, respectively. It is suggested that the number of non-fatal accidents actually reported is rather less than should have been reported; perhaps, by a factor of two!

Cost control systems involve additional overheads. The greater the degree of control the more effectively the direct cost of a project can be controlled, but there is a point at which the total costs are a minimum. Beyond this point, there is no greater saving. Experience suggests that relatively few cost codes are used in the cost control system. It has been reported [5] that recording cost data becomes erroneous if large numbers of codes are used. For example, 30, 200 and 2000 cost headings result in 2%, 50% and 98% of items being misallocated.

Too often, cost control systems are merely cost monitoring systems. They tend to be passive, backward looking systems as opposed to active, forward looking systems in which an attempt is made to take corrective action if cost targets appear to be excessive.

Planning into practice

The construction industry is usually involved on one-off projects; these are invariably managed with a new team. As the location of each project varies widely, the workforce is largely new and the conditions under which the project is undertaken can differ depending on the site conditions, climate, etc. Unlike the manufacturing industry, the approach to a project is rarely uniform. The size of the main contractor's site organization, which is comprised of technical and non-technical staff, often depends on the size of the works. However, as the use of subcontractors has become more popular, the main contractor's team has reduced in size. In the case of a management contractor taking responsibility for the works, all the site work is subcontracted.

After the design of the project has taken place, in whole or part depending on the nature of the project management, it is the responsibility of the contractor to construct and maintain the project in accordance with the contract documents. The prime requirements are fitness for purpose and the maintenance of safety during construction and operation. In terms of the work on the construction site, the essential task is to carry out this work in accordance with the drawings. Traditionally, the contractor has been given a measure of freedom to carry out the project as he believes appropriate and this has been extended to the design and execution of temporary works, that is, works that are necessary in order to satisfactorily construct the project but are not part of the completed project. An example is the use of temporary supporting structures for the erection of the main structure. Indeed, a keen price for a project may reflect the ingenuity of the contractor in providing inexpensive temporary works; it may also reflect excessive optimism and cutting corners. Although the designer might check the contractor's proposals, this in no way relieves the contractor of his responsibilities and liabilities. Some consulting engineers refuse to check erection proposals [6]. However, the contractor should welcome an independent checking system in the interests of the safety of personnel and the works. A notable accident which occurred during bridge construction where erection procedures contributed to the failure was at Milford Haven in 1970 (Figure 11.3). Four men were killed.

The work can be carried out efficiently if the site is laid out in such a way that the temporary buildings, for example, offices, stores and workshops, are conveniently located with respect to the permanent

Figure 12.1   *Planning a confined site in a city centre*

works. See Figure 12.1. This results in an orderly arrangement which facilitates the economy of construction and administration. The construction of a high-rise building on a confined city centre site, for example, requires the efficient storage of materials, offpeak deliveries, high speed vertical travel with site facilities placed to minimize operative travel. At the Canary Wharf tower there were canteen and lavatory facilities on every fifth floor of the tower to avoid the inconvenience and delay that would be caused by excessive descent and ascent of the 600 operatives.

Construction is a commercial venture in which both the promoter and contractor invest and are subjected to financial risks. The promoter will have been apprised of the order of funds which must be made available to meet the contractor's payment certificates should the contract develop according to plan. Contractors may front load the early stages of the construction works to improve cashflow during the mobilization period of the contract. As the contract progresses, should the construction of the works be delayed by unforeseen contingencies and payment to the contractor be delayed or costs increased, the contractor can be subjected to financial difficulties. In addition, the promoter may be financially disadvantaged by delay in the commissioning of the works due to loss of revenue. It is to the advantage of all parties to ensure that difficulties and delays do not occur or their effect is minimized.

References

[1] *Civil Engineer's Reference Book*. Butterworth, London, 1975
[2] Pugsley, A.G. 'The engineering climatology of structural accidents'. *International Conference on Structural Safety and Reliability*. Washington, 1969
[3] Cormican, D. *Construction Management: Planning and Finance*. Construction Press, London, 1985
[4] NEDO. *Faster building for industry*. 1983
[5] Harris, F. and McCaffer, R. *Modern Construction Management*. BSP Professional Books, London, 1989
[6] Blockley, D. *The Nature of Structural Design and Safety*. Ellis Horwood Limited, Chichester, 1980

13

Manpower planning and human relations

One of the problems that dogged construction during the boom years of the middle to late 1980s was skill shortages – the prospect of serious skill shortages when demand picks up looks even more likely now than it did in 1989.

Kick Start. How to get construction going again. Union of Construction, Allied Trades and Technicians, 1992

Manpower and manpower organization

The construction industry is labour intensive and heavily craft based. Employment in the industry fluctuates with the workload. During the downturn in construction in the early 1980s, over 250 000 employees lost their jobs and skilled craft labour left the industry. During the same period, the total number of apprentices and trainees fell by 30%; first year apprentices and trainee recruitment also fell by about 40%. Both these factors contributed to a serious skills shortage during the up-turn in the industry in the late 1980s.

There has been a marked change in the structure of the industry during the 1980s. The number of self-employed people has more than doubled. Over 40% of construction employees are now self-employed. The total labour force of the British construction industry during the past few years is shown in Table 13.1 and Figure 13.1. This is based on the Standard Industrial Classification 1980 for employees [1] together with the current estimates for the self-employed [2] and a recent report on the future skills needs of the construction industry [3]. Data from different sources often show small variations.

Many of the employees are employed by small firms. Since 1976 there has been a sharp increase in the number of small firms brought

Table 13.1 Manpower of the construction industry [1–3]

								Year								
	79	80	81	82	83	84	85	86	87	88	89	90	91	92	93	94
Employees (,000)	1310	1320	1230	1120	1015	1106	1086	1051	1058	1086	1109	1117	1041	923	840	781
Self-employed (,000)	340	310	388	420	409	454	470	488	535	592	696	715	657	597	571	604
Total (,000)	1650	1630	1618	1540	1424	1560	1556	1539	1593	1678	1805	1832	1698	1520	1411	1385

Figure 13.1   *Manpower of the construction industry*

about by the labour being shed by the larger firms finding that the law governing self-employed, introduced in 1978, facilitated labour only subcontracting and self-employment practices. The smaller firms undertake repair and maintenance and act as subcontractors to larger firms.

Because of the large number of self-employed persons in the industry, it is difficult to make a close analysis of the composition of the labour force. However, for 1995, the breakdown shown in Table 13.2 is believed to be valid. The self-employed have been included as working proprietors and other self-employed.

Table 13.2   Composition of the labour force 1995

Directors, working proprietors and senior management	300 000
Office staffs	300 000
Direct labour, local authorities	129 000
Operatives: apprentices	80 000
tradesmen	350 000
labourers, etc.	120 000
Other self-employed	500 000
Total labour, construction industry	1 779 000

Recruitment

Wastage from recruitment, death, accident and illness, and changes to other occupations necessitates the industry finding over 80 000 recruits every year if its numbers are to be maintained. The bulk of these recruits are at the craft and operative level. Many are unskilled in terms of formal training. Also required are technicians and, forming the top tier of the industry, the professional staff. Unfortunately, not only has construction an image of a difficult, demanding and unbecoming occupation with an adversarial culture, but it is not seen as progressive, offering further training or financial advantages. A survey carried out in 1989 by *The Guardian* suggested that only 8% of students 'most liked' construction. This was the lowest percentage for the 'most liked' careers. Thirty-one per cent of students placed construction near the top of the 'least liked' careers.

Schools are not well informed about the construction industry. It is seen to be male-dominated and, particularly at the craft level, unattractive to women and ethnic minorities. Schoolteachers often judge that careers in the construction industry are better suited to low achievers academically. Craft students, for the most part, are school leavers who have not benefited academically from a school career although others have fathers already working in the construction industry and choose to enter craft education. The recruitment to craft and advanced craft courses has risen gradually by 10% over the past 10 years to 1990; in 1989 there were some 70 000 students on courses in further education colleges of which 70% were studying craft level courses. Private training agencies accounted for a further 10–15% of students. The highest enrolments were in carpentry and joinery (20 000), electrical installation (19 000), and brickwork (14 000).

Courses are usually undertaken on a day release or block release mode of attendance. The steady increases in recruitment have been due to more adult and women students entering the courses; they form 20% and 2%, respectively, of the student numbers. The withdrawal rate from craft courses is high and of the students who take their examinations some 30% are unsuccessful. Not all students are lost to the industry. Indeed, firms with high workloads are sometimes responsible for encouraging the better craft students to leave their courses before completion to undertake full-time employment.

Technicians are recruited largely from diploma or higher diploma construction courses. Some recruitment takes place directly from school into trainee technician status and certificates or higher certificates are undertaken on a part-time basis. There are approximately 24 000 students enrolled on certificate and diploma courses of whom 17 000 are on the former. Correspondingly, there are approximately 12 000 students enrolled on higher certificate and higher diploma courses of whom over 9000 are on the former. Most of the higher certificate students, i.e. 6000, are on building courses; 2000 are on civil engineering courses and 1000 on building services courses. Many students undertaking higher certificate or higher diploma courses regard them as a vehicle by which to achieve graduation and eventual professional recognition.

Professional construction staff are recruited largely from degree and postgraduate programmes although, in 1989, there were approximately 6000 students on courses leading to examinations set and marked by professional institutions. There is a range of degrees that result in qualifications which lead to employment in the construction industry. The most popular are architecture, building surveying, estate management and quantity surveying. In 1995, 30 000 students were reading degrees and postgraduate courses leading to construction qualifications. Although withdrawal rates from some of these courses are often high, e.g. 30%, once committed to a construction course most students enter a construction career. After qualifying, over 95% of students enter the construction industry, at least for the first part of their career.

There has been a number of strategies used to better inform school children about the benefits and attractions of a career in the construction industry. These have included careers advice by lead bodies and employers in the industry, project work promoted by the Concrete Society, and visits to construction sites and materials suppliers. One initiative which has been more successful than most, is to introduce vocational courses into schools which lead to GCSE qualifications in building studies and other construction areas. This has been done by means of Building Curriculum centres in which school children can work towards a GCSE examination at the end of a 2-year course. In some cases, over 500 schoolchildren undertake courses in a school year, some from primary schools. There are 40 curriculum centres already established in Britain and a further seven under consideration. The opportunity for school children to undertake 'hands on' activities, rather than listening to presentations and making visits, is seen to demonstrate the attraction of construction work whilst developing group and other transferable skills.

The total number of potential recruits to the industry, based on the number of students undertaking construction education and training, tends to correspond fairly closely to the number of recruits required. However, there is a mismatch between the number of students on some courses and the needs of the industry (see p. 260). This has given rise to unemployment, professionally trained staff undertaking technician duties, unsatisfactory conditions of employment and depressed rates of pay. The stop–go nature of the industry has not facilitated manpower planning.

Skills shortages

More recently, between 1990 and 1992, the training budgets have been cut by £706m. For example, the CITB had its budget for youth training schemes reduced from £79.6m in 1989/90 to £40.3m in 1990/91, and the construction industry has been seriously hit with 40% of potential trainees moving away from construction. The future skills shortage is likely to be severe with a shortfall of 250 000 skilled operatives by the late 1990s [4].

Performance and motivation

The measurement of productivity in the British construction industry and its comparison with that of other countries is a difficult task to undertake as data are not readily available. Productivity can be calculated by finding the output per worker by using the value of the output of the industry divided by the total number of workers; this must include the self-employed workers whose numbers are not accurately known. As the cost of materials is included in this measure of productivity, the escalation of material prices can have the effect of incorrectly enhancing productivity. The distribution of work can also distort the calculations; for example, an increase in the proportion of repair and maintenance in the total output influences the amount of materials used. Nonetheless, this approach is sufficiently accurate to give a first-stage estimate and indicates trends in productivity. Materials can be excluded from the former calculation by using the net gross domestic product in construction divided by the total number of workers. This is a measure of the 'value added' by the industry, that is, the excess of the value of its current output over the value of goods and services purchased from outside the industry and used in production. Table 13.3 and Figure 13.2 have been produced from a number of sources [1, 3, 5]. The data presented are not internally consistent and are used to indicate trends.

The data show that, over the period of time considered, the number of persons employed in the construction industry has varied by almost 20%. The cycle of employment shows similarity to, but lags behind, the work output cycle. Consequently, productivity is poor as the industry enters recession. Since 1979, there has been considerable improvement in productivity. However, reference [6] indicates that up to 45% more manhours are required to complete a project in Britain than are required by other northern European countries. Further afield, international comparisons [6] suggest that 15% and 50% more manhours are required to complete a project in Britain than in Japan and the United States of America, respectively. Based on a gross domestic product index [7], the position of the British construction industry has been extrapolated as deteriorating rapidly when measured against the performance of the United States, Japan, France and Germany. Recent data [3] show that, despite Britain's improved productivity, Germany, France and Italy are 104%, 41% and 12% more productive than Britain, respectively. The reason for poor productivity has been cited as excessive unproductive time on site due to materials delays, lack of equipment, absenteeism; in short, less attention paid to site logistics. For example, the use of power tools and larger items of plant and equipment is more prevalent in the civil engineering sector of the industry and rather less emphatically in other sections of construction. A 1990 study has shown that bricklayers were productive for only 55% of work time [5]. Effective planning and supervision, which determine productive efficiency, depend on the method by which the site is managed and organized. A comparison [8] between British and US contractors undertaking high-rise construction indicated that US fabricators used a 3-day floor cycle compared with a 10–15-day floor cycle in Britain. The fast track system (see Chapter 5) is designed to speed up the construction process by overlapping design and fabrication schedules. Large, shop fabricated components delivered to site for rapid erection minimize the actual construction work. Problems of co-ordination and task complexity are reduced. Although technical innovation is not the com-

Table 13.3 Measurements of British construction productivity relative to 1979

	Year															
	79	80	81	82	83	84	85	86	87	88	89	90	91	92	93	94
Output at 1985 prices £bn	29.18	27.83	25.14	25.47	26.64	29.31	29.66	30.64	34.17	37.46	39.48	39.87	37.17	35.68	35.00	36.13
Employees (million)	1.65	1.63	1.62	1.54	1.42	1.56	1.56	1.54	1.59	1.66	1.81	1.83	1.70	1.52	1.41	1.38
Productivity based on output	1.00	0.97	0.88	0.94	1.06	1.06	1.08	1.12	1.22	1.28	1.23	1.23	1.24	1.33	1.40	1.48
Productivity based on net GDP [1]	1.00	0.96	0.89	1.01	1.16	1.11	1.11	1.17	1.27	1.27	1.28	1.29	1.28	1.38	1.47	na
Productivity [5]	1.00	0.93	0.89	0.93	0.97	0.97	0.99	1.03	na	na	na	na	na	na	na	na
Productivity [3]	1.00	0.97	0.88	0.93	1.03	1.03	1.07	1.13	1.16	12.0	1.17	na	na	na	na	na

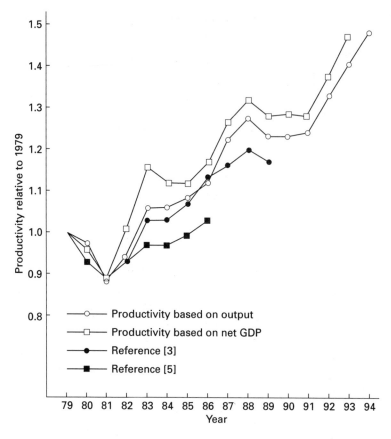

Figure 13.2   *Measurements of British construction productivity relative to 1979*

plete answer, the development of novel components to improve the straightforwardness of construction will, in turn, improve productivity. It is said that in some countries, particularly the USA, there is a productivity culture which results in pressure from all employees to develop and contribute to schemes of work which may result in an improvement in productivity.

Productivity in Britain has been improved over the past decade largely by an increasing use of subcontractors. The main contractor has sometimes only a supervisory team on site. This has increased output but caused a greater fragmentation of the industry. If this continues, the British construction industry may find itself made up largely of subcontractors, with low margins, low profitability and low esteem, in short, the navvies of Europe [9], rather than a market leader of new technology. It has been recognized that a larger number of smaller companies is not conducive to innovation. It is large firms that tend to innovate. Some restructuring is considered to be an important development in order to further improve productivity. Improved productivity will ensure that the industry will be able to recruit a sufficiently large and committed labour force in order to meet its commitments.

Registration in the construction industry

Performance indicators for the construction industry, for example, low productivity, poor product quality, unsatisfactory safety record, imply that a major contributory factor is an inadequately trained and qualified workforce. The increase in self-employment and the decline in training means that the industry is presently facing serious skills shortages which will limit its future effectiveness. To help remedy the situation, an extension of the existing qualification-based registration schemes, which already cover 10% of the operatives, is suggested. This might include both the statutory registration of individual operatives and professionals in addition to companies. Issues of health and safety, and consumer protection could be more easily controlled. Now that the National Vocational Qualifications (NVQs) are coming on-stream, the time might be opportune to introduce registration. Additional benefits would be to dispel the cowboy image of the industry and prevent the losses resulting from abuses in personal taxation which are approaching £1m every week.

Although the system of statutory registration has not yet been widely adopted in the European Union, it does operate in various forms in several countries. Where public safety is an issue, a high degree of regulation exists.

Labour relations

The history of labour relations in the construction industry has been one in which there has been direct confrontation in addition to more subtle techniques for influencing industrial developments. During the past decade or so, the position of the workforce has been weakened due to recession in the industry. Fragmentation brought about by self-employment and the growth of subcontracting have restricted the opportunities available to the main body of the workforce to improve their wages or conditions. These difficulties have not only been confined to the operatives. The technicians and professionals have also seen the worsening of their conditions of service. This has encouraged a migration to more lucrative areas of employment. Large contractors have been pleased not to have to retain a core workforce. Instead, the reduction of the recurrent costs made possible by the smaller workforce has enabled contractors to operate more flexibly [10].

The Working Rule Agreements were among the early instruments fashioned by employers and operatives to try to prevent widespread disputes in the construction industry [11, 12]. Typical of these is the Working Rule Agreement devised by the National Joint Council for the Building Industry. It provides for wages to be decided on a national basis but within the framework there are regional or local variations and additions which are nationally approved. It was agreed in principle in 1920 but it was not until 1932 that the present National Joint Council came into operation. The Council consists of representatives of the Building Employers Federation (formerly National Federation of Building Trades Employers) and the following trades unions:

- National Federation of Roofing Contractors
- Union of Construction, Allied Trades and Technicians
- Transport and General Workers' Union
- General, Municipal, Boilermakers and Allied Trades Union
- Furniture, Timber and Allied Trades Union

The constitution of the National Joint Council empowers it to deal with all aspects of employment both for skilled and unskilled workers but some of its functions are delegated to local joint committees. These are set up by the regional joint committees, wherever appropriate. These committees form the links between the National Joint council and the workers and employers in the localities. For the purpose of settling disputes and difficulties, the regions are required to set up regional conciliation panels which meet as soon as practicable and in any event within 7 days. In the event that a party to a dispute fails to implement a regional decision, the other party or parties shall have the right to appeal to the National Conciliation Panel.

The matters dealt with in the Working Rule Agreement are the following:

Section 1 – National working rules (NWRs)
    NWRs   1–5 Wages
    NWRs   6–13 Hours, conditions and holidays
    NWRs   14–18 Allowances
    NWR   19 Retirement and death benefit
    NWRs   20–21 Apprentices/trainees
    NWR   22 Scaffolders

NWRs   23–24 Safety
NWRs   25–27 General (e.g. grievances, disputes and differences)

Section 2 – Miscellaneous
e.g. incentive schemes, safety and welfare

Section 3 – Regional rules and variations

Section 4 – Associated agreements
e.g. annual holidays with pay, benefits

Two further sections deal with matters such as the memorandum of agreement and the indexing of sick pay.

The Working Rule Agreements, in their various forms, have been used in the construction industry for some decades. The machinery has proved successful and the number of working days lost due to dispute has been lower than most other industries. For example, Table 13.4 and Figure 13.3 show the number of working days lost in construction industry disputes through all stoppages since the late 1970s. This is compared with the working days lost for all industries and services. In total around 600 000 working days were lost in 1993 as a result of labour disputes. This was slightly more than in 1992 when fewer days were lost than in any year since records began. The relatively high losses in 1979 and 1984 were due to the 'winter of discontent' and the support of the miners' strike, respectively. There has been a fall in construction disputes over the time period shown. This reflects the growth of self-employment during the 1980s and the relatively weaker position of employees in a bargaining situation. The disputes in the construction industry are at a far lower level than that for Britain as a whole. In recent years it is an order of magnitude less. It is surprising that the growth of subcontracting during the 1980s has not contributed to greater labour difficulties as main contractors do not now have close control over the site labour relations. Instead, negotiations are carried out by a number of different subcontractors

Table 13.4   Working days lost through disputes per 1 000 workers [1]

| | | | | | | | Year | | | | | | | |
	79	80	81	82	83	84	85	86	87	88	89	90	91	92	93
Construction	103	63	36	29	45	223	34	23	14	11	71	8	4	3	4
All industries and services	92	59	63	73	161	1135	299	90	164	166	182	83	40	30	35

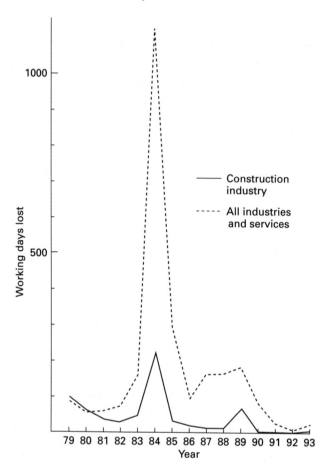

Figure 13.3   *Working days lost through disputes per 1000 workers*

and the labour difficulties of one or two subcontractors could jeopardize the ordered operation of a large construction site. The growth of subcontracting cannot remove the potential conflict between employees of a subcontractor and the main contractor even though the channels through which these relationships are enacted are less well defined.

The trades unions in construction

The largest unions representing the construction industry are:

UCATT: Union of Construction, Allied Trades and Technicians
TGWU: Transport and General Workers Union
GMBATU: General, Municipal, Boilermakers and Allied Trades
           Union
FTAT: Furniture, Timber and Allied Trades Union
EETPU: Electrical, Electronic, Telecommunications and Plumbing
           Union

Table 13.5 shows the variation in union membership of construction workers since 1980; also shown are the number of unions in which construction workers are registered.

Table 13.5   Membership of Trades Unions [13]

	Year							
	80	81	82	83	84	85	86	87
Numbers of Unions	14	14	14	14	14	13	14	12
Membership (,000s)	430	386	353	334	315	311	305	306

By far the largest number of construction workers are registered with the UCATT whose membership in 1987 was 255 000. Most of the rest of the construction workers are registered with the TGWU.

The membership in construction is relatively low, starting the decade with only some 30% of all workers registered. During the 1980s the growth of self-employment and the casual nature of employment contributed to a reduction in membership to less than 20% although the rate of membership decline ceased in 1987. The decline must have contributed to financial losses for the unions concerned and a relative weakening of their position as representing the construction workers. Despite the efforts of the unions during the past few years to attract a greater number of self-employed workers, there has been relatively little success. Yet, unless trades unions do succeed in attracting self-employed workers as members, there is a likelihood that their position will be yet further weakened. This can cause a loss of control of industrial relations by the firms responsible for construction which could be compounded by the wider spread of subcontracting.

Manpower difficulties of the construction industry

When compared with other European countries Britain has the lowest participation rate in education for 17, 18 and 19 year olds [14]. A poorly qualified national workforce detracts from the competitiveness of British products and results in growth-restricting skills shortages. For an industry such as construction which is perceived by prospective recruits as generally unattractive, there is an added danger that the industry will be handicapped by failure to obtain sufficient talent to enable it to operate competitively. Recruits to the industry must be used as effectively as possible with emphasis placed on skills development. These difficulties have been compounded by the recent increases in casual, mobile, labour. Contractors find it to their advantage not to engage large numbers of direct labour employees. This tendency is largely prevalent in building; in civil engineering, for example, the use of casual labour is not more than 30% and in engineering construction it is negligible. Although there have been improvements in productivity using subcontracted labour, these improvements are limited by diminishing returns. The use of casual labour has disadvantages for the building product and for the development of a high technology, productive industry. The immediate disadvantage for the building product is that the main contractor is not in full control of the quality of the product. The self-employed have little loyalty to an employer from whom there are small gains in terms of career and personal development; hence quality suffers. In the longer term there is a serious adverse effect on the levels of training. The training and development of the workforce skills are neglected.

At the graduate and professional level there is a mismatch in the distribution of skills of those entering the industry and the nature of the product [15]. Typically, some 45% of graduates are civil and structural engineers, 20% are architects, 20% are surveyors, 10% are builders and 5% are building services engineers. These skills do not reflect the cost profiles of the construction product. There are few graduate managers compared to other European countries; less than one third of other European countries and Japan. Yet sustained improvements depend on better techniques, better designs, better products and higher quality.

In short, the British construction industry is facing marginalization by better trained and managed international workforces. The manpower difficulties of the British industry place it in a position where it is unable to produce the high quality, efficiently produced products that will enable it to compete effectively.

References

[1] Central Statistical Office. *Annual Abstract of Statistics.* HMSO
[2] Department of the Environment. *Housing and construction statistics.* HMSO. London
[3] *Future skills needs of the construction industry.* A report prepared for the Employment Department. IPRA, 1991
[4] *Kick Start. How to get construction going again.* Union of Construction, Allied trades and Technicians, 1992
[5] *Building Towards 2001.* Building Employers Federation, 1990
[6] Koehn, E. 'International Labour Productivity Factors' in *Journal of Construction Engineering and Management. ASCE,* **112**, 1986
[7] Arditi, D. 'Construction Productivity Improvement' in *Journal of Construction Engineering and Management. ASCE,* **111**, 1985
[8] *UK and US Construction Industries. A comparison of design and contract procedures.* RICS. 1979
[9] Foster, C. *Delivering the people to deliver the goods. WHOM DO WE SERVE? The future of professional education and training in construction.* Construction Industry Council, London, 1992
[10] Ball, M. *Rebuilding Construction: Economic Change and the British Construction Industry.* Routledge, London, 1988
[11] Working Rule Agreement. National Joint Council for the Building Industry, London, 1991
[12] Working Rule Agreement. Civil Engineering Construction Conciliation Board for Great Britain, London, 1991
[13] Waddington, J. 'Trade Union Membership, 1980–87: unemployment and restructuring' in *British Journal of Industrial Relations,* June, 1992
[14] Organisation for Economic Co-operation and Development. 1991
[15] Degrees in building management: demand, provision and promotion Report and Recommendations from the Joint Committee on Higher Education in Building, (Lighthill Report), Building Employers Confederation and the Chartered Institute of Building. 1986

14

Education and training

There are two aspects of education. One regards the individual human mind as a vessel, of varying capacity, into which is poured as much as it will hold of the knowledge and experience. The second regards the human mind as a fire that has to be set alight and blown with the divine wind.

Lord Crowther, 1969

Introduction

British education aims to develop fully the abilities of individuals both for their own benefit and for society as a whole. Much of the post-compulsory (i.e. post-school for pupils over 16 years of age) education is organized flexibly in order to offer a wide range of opportunities for academic and vocational education and to provide for continued study and development throughout life. During the past two decades further and higher education, the post-compulsory sector, has been characterized by change with a considerable growth in the provision of courses, accessability, responsiveness, student numbers and levels of achievements.

Attempts to improve the quality and breadth of education have resulted from the Education Reform Act (1988), the two Government White Papers, *Education and Training for the 21st century* (1991) [1] and *Higher Education : a New Framework* (1991) [2] and the subsequent Further and Higher Education Act (1992). These have identified changes in the way that colleges and universities will be managed, and the move towards student-related tuition fees to emphasize market forces funding. There is a commitment to maintaining a high quality, cost-effective education system which helps to satisfy the nation's economic and social needs and at the same time offering opportunities to individuals in the advancement of their knowledge and the pursuit of scholarship.

Definition

Education and training are terms which are synonymous. Education is a systematic course of instruction which is broadly based to develop the mind. It may be academic but it can also be vocational. Training, on the other hand, is much more vocationally orientated to suit a particular work-related occupation. This may be specific and relate only to a single industry or it may be designed to incorporate transferable skills for other areas of work. Training and education are a combination of knowledge, understanding, skills and application. They aim to change an individual's attitude and improve his or her performance.

Demography

In 1980 there was a peak of over 1 million 16 year olds [3]. However, due to the demographic decline resulting from the fewer births in the late 1960s and 1970s, this number will fall to below 700 000 by the mid-1990s reaching a trough in 1993, before rising marginally through to the end of the decade. The decline represents a reduction of 30% nationally although it is more severe in some regions of the country than in others. It is less significant amongst the socio-economic classes I and II where the decline is between 15% and 20%. Amongst the classes III to IV the reduction will be about 50%. The significance of this phenomenon for the construction industry and its education provision may be considerable.

1 Construction courses in higher education [4] which at the present time are continuing to expand, will need to be marketed more aggressively as the competition for student numbers increases.
2 The construction industry is often not a first choice for students, and is seen by some as an unattractive industry, with limited opportunities and career prospects, at the mercy of the British weather, severely influenced by government policies and subject to a high number of bankruptcies and insolvencies.
3 An upturn in the economy will encourage some students to opt for jobs in the construction industry rather than to undertake education and training.
4 Building craft courses, which typically recruit from socio-economic classes III and IV, will face further problems of recruiting

students in a recession, due to lack of training opportunities and in times of boom due to other opportunities in other occupations and industries.

The recession in the country and the lack of job opportunities has encouraged more school leavers to study full-time in either further or higher education. Whilst there has been some expansion of built environment courses, overall whilst student numbers were increasing in many other subject areas, student numbers in the built environment having been declining throughout the early 1990s. The relatively poor image and lack of job and career opportunities in the construction industry have helped to discourage potential applicants from these courses. Consequently any upturn in construction activities that might be forecast towards the end of the century is likely to result in a dearth of skilled manpower at all levels of activity.

Qualifications

Approximately 14% of all pupils leave school with two A-levels certificates (Table 14.1). Of these, about 85% enter full-time higher education with the remainder seeking some form of employment. In addition, courses in higher education recruit students with other qualifications such as those of Business and Technology Education Council (BTEC), or enrol mature students on the basis of their experience. Evidence at the present time suggests that educational achievements, which have continued to rise since the 1950s, have now begun to level out [5].

A popular view for the poor performance of school leavers is that high failure rates are necessary to maintain high standards [5]. The

Table 14.1   Qualifications of school leavers 1983/84 and 1993/94

| | 1983/84 | | 1993/94 | |
	%	Cumulative	%	Cumulative
2+ A levels	13	13	21	21
1 A level	4	17	8	29
5+ GCSE	8	25	24	53
1–4 GCSE	25	50	40	93
No qualifications	50	100	7	100

true effectiveness of the above, which rejects almost 80% of its products at the highest level, serves neither social justice or the economy. It is dominated by an under-expectation of what individuals can achieve, resulting in lower levels of attainment and narrower competencies. The education system has bowed to the needs of the more academic, leaving the potential of the majority as an untapped natural resource. There is a tendency to undervalue skills and professionalism. Vocational training has favoured lower level qualifications for traditional jobs over high level skills for newer industries. The position is further aggravated by both the failure and withdrawal rates of students who enter higher education. Less than 80% of these successfully complete their construction courses. Implications for the construction industry are that amongst a declining supply of young people the construction industry may be forced to recruit less-well-qualified individuals than previously. This may require a reappraisal of what can be taught in the time which is available.

Table 14.2   Highest qualifications of 16–59 age group 1995

NVQ Level 5	Higher degree		2
NVQ Level 4	Degree/Diploma		
	First degree	8	
	HND/HNC	4	
	Teaching	1	
	Other	5	18
NVQ Level 3	Vocational	10	
	Advanced GNVQ	6	16
NVQ Level 4	Vocational	10	
	Intermediate GNVQ	11	21
NVQ Level 5	Vocational	6	
	Foundation GNVQ	16	22
None			21
			100

Table 14.2 shows the NVQ levels of qualification and how these were distributed throughout the 16–59 age range in Great Britain in 1995. In the 1950s, approximately 2% of the population obtained a university education. This has now risen to about 35%. Approximately 2% of the population currently have a higher degree. It is expected therefore that the level of qualifications in Britain will rise considerably during the next ten years. For example, building craft qualifications a few

years ago would allow an individual to progress within a company. Today membership of a professional institution is now frequently a prerequisite with the possession of a degree becoming more common.

Education and employment

There has been a large increase in the number of pupils and students staying on in full-time education after completing their mandatory education at 16 years of age (Table 14.3). This is due, in part, to the greater recognition of the importance of acquiring formal qualifications, reduced opportunities for employment due to the recession of the early 1990s, and to the demographic decline amongst those who did not traditionally stay on at school or college [5]. The staying-on in full-time education rates vary across Britain from as few as 35% to as many as 80%. By 1991 this figure had reached an average of over 50% and by 1970 in excess of 70%. Fewer students are, for example, joining Government Training Schemes due both to lack of student placements and to other opportunities elsewhere. This trend may have serious implications for building craft training where 20% of the students are placed, and which represent about 80% of the building craft training programme.

International comparisons

Compared with the other industrialized countries around the world, Britain fares badly [7, 8] in terms of staying-on rates of school leavers (Table 14.4). Whilst the figure continues to increase, it still lags behind Britain's competitors which have also improved their performance in this area. Assuming that the construction industry draws people from the same strata of society internationally, then these statistics make gloomy reading, considering the implications of the Single European Market in 1992, the importation of Japanese firms into Britain and competition for worldwide markets.

Colleges

The majority of the formal education and training of craft, technicians and graduates for the construction industry in Britain takes place in further and higher education institutions. These include 500 colleges of further education (FE) and 125 institutions in higher education (HE). In 1992 the polytechnics became universities with the Polytechnics and Colleges Funding Council (PCFC) and the Universities Funding Council (UFC) joining to form a single body

Table 14.3   Percentages of 16 year olds [6]

	1975	1980	1985	1988	1990	1991	1992	1993	1994
School	22	23	25	24	30	42	46	40	38
Further education (FT)	10	10	12	13	17	19	23	32	33
Youth training schemes	–	5	18	25	20	11	7	12	11
In employment	63	54	30	28	33	28	24	16	18
Unemployed	5	8	15	10	–	–	–	–	–
Totals	100	100	100	100	100	100	100	100	100

Table 14.4   Full-time participation in education and training – international comparisons [6, 9]

	16-year olds		17-year-olds	
	1988	1992	1988	1992
Britain	50	75	35	57
France	82	92	73	89
Germany	71	95	49	93
Netherlands	93	97	77	92
Japan	93	95	84	90
USA	94	94	87	95

named the Higher Education Funding Council for England (HEFCE). At the same time further education colleges were to be managed under a new Further Education Funding Council (FEFC). Similar unified funding councils were established for Scotland and Wales. An additional 10% of building craft training is undertaken by private training companies. These are outside of the college system although the students aim to achieve the same national vocational qualifications. Other courses of study are also provided through, for example, distance learning, whereby the students study at their own pace at home and have assignments marked by tutors.

Further Education

About 250 of the colleges of further education in Britain provide courses for school leavers who want to follow a career in the construction industry. All of these colleges offer courses which are of a vocational nature and many provide courses for the different building crafts. The range of crafts (and the equivalent of advanced craft certificates) varies, but typically include brickwork, carpentry and joinery, painting and decorating, plastering and plumbing. These are known as the major crafts. In the larger departments several other crafts are also taught such as shop-fitting, roadworks, floor, wall and roof tiling, etc. Building craft courses are usually studied on either a part-time or block release basis, although some courses are now being designed as full-time study. The courses take about two years to complete up to craft certificate level (National Vocational Qualifications (NVQ) level II) and a further year for the advanced craft certificate (NVQ level III). The courses have been designed within a national framework to meet industry's needs and often reflect a flavour of local industry. The principal awarding bodies are the City and Guilds of London Institute (CGLI) and the Construction Industry Training Board (CITB), under the umbrella of NCVQ. The CITB is the construction industry's lead body for craft education. There are currently about 70 000 students on building craft courses in further education and a further 7 000 in training elsewhere. Some of the latter are mature students on adult training courses.

Further education colleges also provide courses for students who wish to follow a non-craft based career in the construction industry. Courses are provided for building and civil engineering technicians leading to an award of a GNVQ (General National Vocational Qualification). These courses have replaced the former National and Certificate and National Diploma courses and are designated advanced GNVQs. GNVQs may also be offered at intermediate level and these are normally offered in secondary education. There are relatively few of these aimed at the construction industry, due to a lack of student demand and teacher capability. These courses typically take two years to complete and are the equivalent of A-level qualifications. They can thus be used as an alternative for entry on to courses in universities.

A majority of students having completed an advanced GNVQ usually enter higher education on an undergraduate degree programme (full-time or part-time) or an HND or HNC which is validated by BTEC. Those who choose to study on a part-time basis do so whilst in appropriate employment. An increasing proportion of FE colleges now offer HNC or HND programmes, sometimes in conjunction with their local university.

Progression

Education for the construction industry includes ladders of opportunities to enable individuals to achieve their optimum level of attainment. Table 14.5 illustrates diagrammatically the options which are available depending upon the qualifications obtained from school.

Higher Education

The universities (which include the polytechnics, designated as universities in 1992) provide courses in higher education (HE). Some of these institutions also franchise to further education, parts of their courses. The range of provision offered in HE includes higher technician, undergraduate courses for BA and BSc degrees, higher degrees MSc, MPhil and PhD, and continuing professional development

Table 14.5   Ladder of progression

School	Further Education	Higher Education	Employment
less than 4 GCSEs- —GNVQ Intermediate			
—Craft			Craft
4 GCSEs	—GNVQ Advanced—HNC		
5 GCSEs		—HND	Technician
1 A-level		Degree	Profession
2 A-levels		Post graduate	CPD for all

(CPD) for those already in employment. About 70 universities offer undergraduate courses for at least one of the built environment professions.

The older university provision was dominated by civil engineering courses and these, in addition to courses in architecture, surveying, planning and building, are now provided within the new university sector (Table 14.6). There are currently over 200 full-time or sandwich degree courses in the built environment. Over 50% of these courses are in the former polytechnics. Over 30% are in one of the surveying disciplines and 25% are civil engineering degrees. Almost 8500 students were on undergraduate courses in the built environment in the universities compared with 14 500 in the former polytechnics. The latter figure excludes the over 6000 students who were studying for

Table 14.6   First degrees and diploma courses in the built environment [6]

	Full-time sandwich	Part-time	HND	Totals
Architecture	34	4	1	39
Building	26	12	23	61
Civil Engineering	55	9	20	84
Surveying	75	30	10	115
Others*	52	5	6	63
Totals	242	60	60	362

* Building services engineering, landscape, planning, environmental health and housing

degrees on a part-time basis. Courses are continuing to be developed to meet the demands and changing needs of industry and commerce and the preferences of students.

The older university undergraduate courses are typically of 3 years duration and full-time. By comparison many of the new universities' courses are of 4 years duration and include a sandwich element component of industrial or commercial training. This requires students to undertake supervised work experience before completing the final year of the course. The 'thick' sandwich arrangement is more common where students are in employment for at least 48 weeks in their year of placement. The 'thin' sandwich often allows for two separate 6 months of work experience, and has the advantage of providing contrasting employment situations where this can be arranged. During the 1980s there was a substantial growth in the availability of degree courses many of which could be studied on a part-time basis. Some of these replaced the former professional body courses. Part-time degrees extend over 5 years concurrently with employment in practice. Some courses are a combination of full-time and part-time study.

Degree courses are aimed at the route towards membership of one of the construction professions. The older universities have concentrated their attention on undergraduate teaching, research and the development of post-graduate courses. The newer universities also offer other types of courses, for example, those for higher technicians. The BTEC Higher National Diplomas (HND) and Higher National Certificates (HNC) are courses of 2 years duration for students who might not have the entry qualifications in the first instance to study for a degree. These typically recruit about 13 500 students annually. The majority of the construction HNDs are in building (23), civil engineering (20) and surveying (10). Some institutions are designing 1-year full-time courses to convert an HND award into an unclassified degree. These arrangements allow students ladders of opportunity to enable those who are sufficiently able, or hard working, to achieve their potential.

Construction courses have introduced aspects of the Single European Market into the curriculum. In some courses students are encouraged to develop language competency to widen their opportunities in mainland Europe. Other courses are emphasizing the varying nature and cultures of the construction industry in different EU countries. In addition, many students now have the opportunity to study or visit an overseas country as a part of their course.

In some of the larger faculties of the built environment, students of different professional disciplines are able to study subjects together

Table 14.7 Construction students in further education [6]

	Students
Craft courses	70 000
Advanced craft courses	30 000
Technician courses*	32 000
Other	5 000
Total	137 000

* About 4000 students in FE colleges are on courses which are designated as higher education. These have been excluded from the above and are included with the HE student numbers below. There are almost 200 000 students (Tables 14.7 and 14.8)

Table 14.8   Built environment students in higher education [6]

	Full-time	Part-time	Total HEFC
Higher National Certificates	—	10 400	10 400
Higher National Diplomas	3 000	—	3 000
Professional courses	—	5 500	5 500
Degree courses	25 300	6 000	31 300
Post graduate/other	3 500	2 500	6 000
Totals	31 800	24 400	56 200

during their first year and to work together as project teams at different stages throughout their courses. The integration of the students in this way is seen as beneficial in attempting to break down the discrete professional barriers.

Students

There are almost 200 000 students at any one time studying on the different courses for the construction industry or the built environment.

These, on average, represent about 8% of full-time equivalent students in the different educational institutions.

There are about 20% of 18 year olds currently on full-time or sandwich courses in higher education in Britain. The majority of these who are on construction courses are on degree courses (90%). Other students have chosen to study on part-time courses leading to first degrees, professional qualifications or HNC awards in construction disciplines. Recruitment to all of these courses is generally good with many courses in high demand with some over-subscribed by appropriately qualified students. These courses, as new courses are developed, face increasing competition from other sectors and interests and also from employers for suitably qualified or capable students. The number of students on degree courses in construction registered with the former Council for National Academic Awards (CNAA) increased from less than 1000 in 1965 to over 20 000 by the early 1990s.

A relatively small proportion of students connected with the construction industry undertake post-graduate study for formal qualifications other than for professional qualification. Architectural students are required to obtain a graduate diploma as a prerequisite to membership of the RIBA and for registration with ARCUK. These courses take 2 years to complete which, in addition to 2 years approved training, gives architecture the largest amount of formal study amongst the construction professions (Table 14.9). There is a small growth in the number of higher degree courses, particularly in construction management. Other postgraduate programmes have been developed in conservation studies, information technology, geotechnics, transportation studies, etc.

Most of the professions require students to pass a test of professional competence or final professional examination, prior to being admitted to full professional membership.

Table 14.9 Minimum qualifying periods of the professions

	Degree	Professional diploma	Qualifying experience	Total number of years
Architect	3	2	2	7
Builder	3	–	3	6
Civil Engineer	3	–	4	7
Surveyor (QS)	3	–	3	6
Surveyor (BS)	3	–	2	5

Construction Industry Training Board

The Construction Industry Training Board (CITB) was established by an Act of Parliament in 1964 to improve the quality of training, improve the facilities available for training and help to provide enough trained people for the construction industry. It is partially funded through government, but most of its income is derived from a levy system on contractors based upon the number of their employees. In addition to being appointed as the primary managing agent for the construction industry's Youth Training Scheme, the CITB also provides for a range of skills training at its national training centres. Whilst the emphasis of its activities are on practical skills training it also offers courses in supervisory management. Until a few years ago all of the construction crafts were managed through the CITB, but the electrical and mechanical trades chose, in the late 1980s, to move away from the CITB influence. In a typical year there are about 30 000 students on the CITB programmes, although this number declined by about 50% in the early 1990s due to the recession in the industry. In addition, other students are managed by different training organizations. The CITB has no direct involvement with adult training, although it does provide subsidies for other organizations to undertake this training. In the early 1990s about 6500 adults were on training programmes of this type.

Construction Industry Standing Conference

The Construction Industry Standing Conference (CISC) was established in 1990 to produce a framework of qualifications for the NCVQ accreditation covering technical, professional and the managerial occupations within the construction industry. The CISC's membership is drawn from the chartered and non-chartered professional bodies, the awarding bodies such as the BTEC, and employer, business and trade organizations. There are strong links with further and higher education. An early task of the CISC was to prepare an occupational map of the construction industry in order to identify the tasks and responsibilities performed by the different professional disciplines. This was published in 1991. The second objective was to develop standards by analysing the different job roles into elements of competence. These are grouped into units and will form the basis of the NVQs. It is intended to prepare instruments for assessing the standards for accreditation purposes and finally a strategy for implementing and maintaining the NVQ system. The merits of this approach are:

- development by the industry
- nationally agreed
- relevant
- accessible
- transferable

The NCVQ encourages career progression and will hopefully increase skill levels and quality of performance.

National Council for Vocational Qualifications

The National Council for Vocational Qualifications was established in 1986 to reform and rationalize the structure of existing vocational qualifications. The Council is instituting a system of accreditation for qualifications awarded for the various levels of occupational competence through approved bodies. The competencies relate to standards and, as such, they are not limited by a time constraint. Students will also be able to study by different methods and be able to accumulate credits towards an NVQ. In the construction industry the lead body, at craft level, responsible for NVQ developments is the CITB, and at higher levels the CISC. The new framework will incorporate qualifications that meet agreed criteria within a simple structure of levels. The first accreditations within the NVQ framework were introduced in 1987 and the first four levels are hoped to be operational by 1994. Construction craft courses introduced the NVQ framework in September 1991 [10].

Training and Enterprise Councils

The Government announced its intention to set up a network of Training and Enterprise Councils (TECs) in March 1989. The purpose of these is to restructure Britain's approach to training and enterprise development. They offer employers, in partnership with the broader community, an opportunity to reskill the workforce and stimulate business growth. A network of about 80 employer-led councils have been established in England and Wales. The TECs have executive responsibility for almost £3bn of public expenditure and directly employ about 5000 staff to manage day to day operations. Each council is an independent company operating under a performance contract with the government. They embody five major principles for reform:

- Local: national training and enterprise programmes need to be tailored to suit local markets.
- Employer-led: pursuance of benefits for the whole community.

- Focused approach: to secure the broader aims of community revitalization and the best co-ordination of policies and programmes.
- Performance: attaining greater value for money, improving efficiency and a higher return on investment.
- Enterprise: capable of driving radical reform and with a bold vision that stretches beyond existing programmes.

Confederation of British Industry

Construction education has been alleged, in some cases, to be too narrowly focused and insular from other activities and industries. In 1988 the Confederation of British industry (CBI) published a document titled 'Towards a Skills Revolution' [11]. This document highlighted the need for improved and continuing training in order to improve productivity and efficiency and to become more competitive in world markets. It provides examples of companies improving their performance and profitability through improved training for their workforces. As a part of its recommendations the report suggests that the following are essential core elements which should be present in any education and training:

- Values and integrity: strive for quality when dealing with other individuals and in undertaking tasks and recognize the importance and care and treatment of others.
- Effective communication: use effectively a broad range of forms of communication depending upon the purposes and intended recipients.
- Numeracy: understand, interpret and use effectively numerical information; use numerical based approaches to solve problems.
- Information technology: understand the role and potential of information technology in both learning and work situations.
- Work and the world: have an awareness of the basic economic ideas applied to work such as wealth creation, supply and demand, growth, enterprise, competition, etc., as they apply to individuals, organizations, industry and society at large.
- Personal and interpersonal skills: review one's own strengths and weaknesses and set targets for personal development.
- Problem solving: formulate solutions and devise strategies to organize a reasonable approach to solve problems.
- Positive attitude towards change: be critically aware of the effects of change to individuals, organizations and industry.

Continuing Education and Training

It used to be the pattern that once the final examinations were passed and an individual began to practice in a trade or profession then that was the completion of formal education. It was accepted that through practice there was a continuous developmental process, but this was *ad hoc* and was lacking in objectives. Those who were dedicated sought their own personal development through further study or the acquisition of additional qualifications. Training programmes or formal staff development were largely ignored by both employers and employees. However, many of the professional bodies in the construction industry now recognize the importance of continuing professional development (CPD) and that, if this is to be effective for the profession and the individual, then some form of registration is required. This requirement ensures that the benefits of institutional membership can only be enjoyed if an adequate amount of CPD is undertaken. The benefits of this include higher productivity and profitability, lower staff turnover and absenteeism, innovation, improved customer service, higher quality and improved job satisfaction. Continuing education and training should arise from the present and potential performance of an individual, which aims through enhancing skills, imparting knowledge or changing attitudes, to improve performance of the individual and as a result, that of the organization. It should provide benefits all round.

References

[1] Department For Education. *Education and Training for the 21st century.* 1991
[2] Department For Education. *Higher Education: a new framework.* 1991
[3] National Economic Development Office. *Young People and the Labour Market.* 1988
[4] Department of Education and Science. *Higher Education in the Polytechnics and Colleges: Construction.* 1990
[5] Department for Education and Employment. *Labour Market and Skill Trends.* 1995
[6] Department For Education. *Statistical Reviews.*
[7] Smithers, A. *Beyond Compulsory Schooling.* The Council for Industry and Higher Education. 1991
[8] Spencer, D. 'UK bottom of staying-on league table' in *Times Educational Supplement*, 25 July 1991

[9} *Unfinished Business. Full-time educational courses for 16–19 year olds.* Office for Standards in Education Audit Commission, 1993

[10] Hurrell M. 'Crucial year for skills debate' in *Construction News*, 13 September 1990

[11] Towards a skills revolution. Confederation of British Industry. 1988

15

Innovation and research

... the stronger the R&D effort of a sector, the better its image; even in a fragmented sector. Look at the image of doctors!

Centre Scientific et Technique du Batiment, 1990

Introduction

Roads, bridges and buildings are the mark of civilization and their development can be traced in terms of the innovations that have taken place over the centuries. The major buildings and constructions of earlier civilizations were bounded by the beam and the column. Despite their mental agility, the early Greeks did not conceive the arch. The arch is an engineering concept which belonged to a more practical era. It enables separate masonry blocks to form a structure to span 60 m or so. Joints in compression provide few problems compared to joints in tension. The semi-circular arch used by the Romans and Moors was used in aqueducts and great mosques. The dome, an arch in rotation, provided a clear roofed area which, combined with supporting arches, was not surpassed in size until the modern railway station roofed by means of steel girders. The later Gothic arch was used in the cathedrals of northern Europe. These massive structures were constructed from masonry blocks in such a way that the materials remained in compression despite a labyrinth of aisles and vaultings. The invention of the flying buttress which helped to support the thrust from the building allowed vaults of some 40 m in height. All this was achieved before the mathematics which derived a quantative knowledge of the structural action was applied. The skill of the builders was learned by experience which conveyed a physical understanding highlighting 'know-how' rather than 'know-why'.

Their concepts were extrapolated to encompass larger and larger structures. Not surprisingly, physical and geometric limitations were breached and disasters occurred. After all, there were few statutory and administrative controls and, no doubt, the building owners encouraged innovation which lifted their structures above the commonplace. At Beauvais cathedral several failures took place. This marked the limitations of what could be achieved with the natural material, at least for the shapes that were being constructed [1]. Builders now needed new materials for major structures such as prestigious buildings and bridges. Understandably, there developed a greater concern for public safety and its regulations.

Conservatism in construction

During the middle part of the nineteenth century, an early form of reinforced concrete was devised. Patents were taken out which proposed its use for a variety of structural forms including beams, floors and bridges. Early success was based on sound judgement but many failures took place. Confidence was developed due to a better knowledge of behaviour of the material due to analysis and experiment. The Hennebique system was introduced into the UK at the beginning of the twentieth century and was used for a number of public buildings including the GPO at St Martins-le-Grand. Nonetheless, there was a good deal of opposition to innovation by local authority byelaws and conservatism, allegedly based on public safety, prevented the quicker spread of new ideas.

Conservatism is found in another large sector of construction, i.e. the housing market. The building societies' view of a house is that it must be a good security for a loan. In the case of default by the borrower, the society will wish to realize the security. Hence, the house must be a type which is in constant demand. This means that it should be of traditional construction. Raising a loan for a house which is to an individual specification is more difficult for a borrower. Potential purchasers are unwilling to buy a house which may not be marketable. Traditional structures which are built *in situ* using conventional materials are appealing to the many small businesses that specialize in housebuilding. Problems are more likely to occur in individually designed houses. Builders merchants, too, hold large stocks of conventional materials and are reluctant to carry stocks for which there may be little demand.

An architect is placed in a difficult position. He cannot recommend the introduction of untried and untested methods at the expense of his client; not least, it might result in a claim against him. He has also to consider the limitations of the builder. Can the builder carry out construction work which might require special skill? The architect is interested in inventions that are technically sound and which will result in economies in construction. Sometimes builders have patented their own construction techniques which are based on well proven procedures. However, the construction industry has relied on the R&D which is carried out by publicly funded organizations such as the Building Research Establishment (BRE).

Innovation in Britain

Technical change is accelerating and progressive businesses tend to adopt quickly new techniques and applications. R&D is inseparable from the well-being and prosperity of a country; as it is to the separate businesses within the country. To make direct comparisons between countries' respective inventive strengths it is usual to concentrate on the USA patent statistics which represent a 'level playing field'. Comparisons suggest that a dramatic decline has taken place in Britain's innovative performance since the Second World War when compared with that of other developed countries [2]. The relationship between the per capita expenditure on industrial R&D and the per capita volume of patenting activity has been shown to be statistically significant. Table 15.1 and Figure 15.1 show the trends in per capita patenting over the past few decades. The relative strengths of Western Germany and Japan are clear.

Table 15.1   US patents per million population [3]

| | Years | | | | |
	63–68	69–73	74–78	79–83	84–88
United Kingdom	44.37	57.23	52.04	38.58	44.06
France	26.64	40.10	43.75	35.90	44.56
Western Germany	55.31	85.13	96.56	88.91	113.90
Western Europe	36.70	53.23	56.49	47.60	58.42
Japan	10.40	36.67	56.62	64.11	114.62

The unfavourable situation is compounded by Britain's reluctance to adopt innovative techniques developed elsewhere [4]. Even countries with a greater innovative activity than Britain have a technological balance of payments deficit. This indicates a willingness to take advantage of R&D undertaken elsewhere in the world. There is less

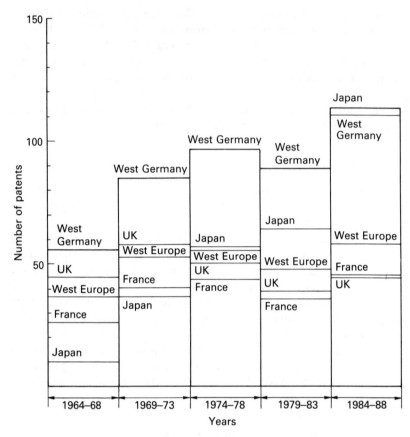

Figure 15.1   *US patents per million population*

willingness to do so in Britain. Overcoming this reluctance is important if best practices are to be adopted. The advantages of innovation are incontrovertible: the production of durable, functional and cost efficient products linked with improved productivity. The former are essential for the British construction industry to survive in overseas markets and to meet the increasing stringent demands of the home market against increasing competition from overseas. Improved productivity means that the construction industry will attract an appropriate share of a scarce labour force as it will pay the highest wages and offer the best terms of employment.

Table 15.2 shows the total number of British patent applications and patents granted by the UK Patent Office over recent years. There has been a substantial decline in the numbers of applications and

Table 15.2 Patents applied for and patents granted by UK Patent Office [5]

		Year										
		80	81	82	83	84	85	86	87	88	89	90
Total: all activities	A	na	na	na	na	na	19 117	15 884	15 354	14 067	13 765	12 695
	G	na	na	na	na	na	20 880	16 206	13 049	11 458	10 138	9 396
Construction	A	na	na	1 083	934	983	1 006	879	749	790	748	733
	G	na	na	1 008	999	694	906	838	682	671	320	382
% Construction	A	na	na	na	na	na	5.3	5.5	4.9	5.6	5.4	5.8
	G	na	na	na	na	na	4.3	5.2	5.2	5.9	3.2	4.1
Construction as % of GDP		6.2	5.9	5.9	6.0	6.0	5.8	6.2	6.4	7.2	7.4	7.6

A = Applications; G = Grants

patents granted due to greater activity in the European Patent Office. The European system has resulted in a decline in the numbers of applications filed directly in the European member countries. Nonetheless, in terms of the contribution of the construction industry to the GDP, i.e. 7.0%, the inventive activity is low and, when compared with earlier years on a relative basis, is declining.

Table 15.3 and Figure 15.2 show the comparison between the construction industries' patenting efforts for several developed countries based on patents granted by the US Patent Office. The British activity has failed to match the pace of others during the decade and of the main European competitors, only Italy is less active. Germany sealed over twice as many patents as Britain.

Japan, which was only a little more active than Britain in 1981, sealed over twice the number of patents as Britain in 1990.

The number of patents does not give a true picture of the construction industry's innovation, for the results of some of the research work are made generally available to the industry whilst, of the rest, a great deal is kept secret and used only in a firm's own contracts. For example, the statistics cannot pick up all the R&D work that takes place on site to find innovative solutions to construction problems. Nonetheless, it serves as one relative measure against major overseas competitors. A further measure is the expenditure of the construction industry on R&D.

Table 15.4 shows that expenditure on R&D has increased during the last decade or so in current prices but in real terms has fallen. Presently it still amounts to 0.48% of the construction output. Construction companies contribute 10% to this sum which is approximately one-third of that of their competitors in France, Germany and Italy. By way of comparison, the Kajima Corporation in Japan produces a wide range of high technology products and services designed to accelerate the overall innovation of construction systems. In respect of expenditure on R&D, the British construction industry lags far behind both competitors overseas and other British industries. For example, all British industries spend some 2.3% of turnover on R&D. How much it is fit to spend on R&D is a matter for conjecture and depends on how this work is selected and organized. However, it is believed that R&D funding is low considering the economic importance of the industry [7, 11]. British companies see progress in construction R&D as largely 'techniques' improvements as distinct from 'products'. They are rarely involved in patenting. Although the public sector accounts for only 25% of all construction in Britain it is often believed that public sector funds should pay for a substantial portion of construction R&D.

A comparison of the relative strength of British manufacturers and

Table 15.3 Construction patents granted by US Patent Office [6]

Country	Year										
	80	81	82	83	84	85	86	87	88	89	90
France	na	56	40	52	74	64	61	86	60	88	72
West Germany	na	123	105	106	124	118	156	187	143	180	154
Italy	na	24	21	14	28	30	24	45	16	34	33
Japan	na	72	77	77	97	105	133	175	129	163	193
GB	na	62	46	44	58	61	60	82	60	76	72
US	na	1178	964	928	1212	1114	1297	1525	1250	1716	1537

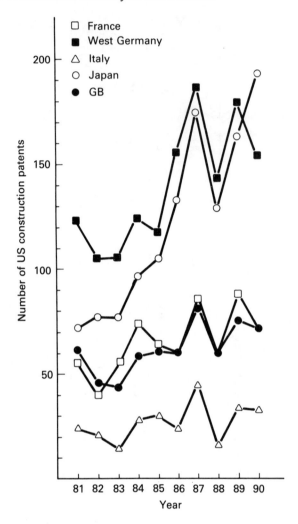

Figure 15.2   *Construction patents granted by US Patent Office [6]*

their mainland European counterparts in meeting architectural specifications when producing building components led to a number of
key issues [12]:

- British manufacturers are reluctant to respond to requests for non-
  standard products. Little regard is paid to the potential value of
  contributing to innovation.
- British manufacturers do not invest in new plant necessary to

Table 15.4 Expenditure of the British construction industry on R&D [7–10]

	Year					
	81	84	87	90	92	94
Construction output (£ ,000m)	21.5	26.2	34.6	55.3	47.5	49.4
R&D (£m)	153	146	246	198	215	236
%R&D Overall	0.71	0.56	0.71	0.36	0.45	0.48
%R&D Manufacturers/ private sector	0.42	0.31	0.39	0.25	0.29	0.28
%R&D Public sector	0.23	0.22	0.25	0.12	0.17	0.19
%R&D Contractors/ consultants	0.06	0.03	0.07	0.05	0.05	0.04

produce components to new specifications. Decisions are controlled by short-term financial considerations.
* Sufficient R&D funds are not available to design and develop new construction components. Few new products are currently being developed by the British building industry.

Overall, it was judged that there is not a sufficient awareness among British contractors and manufacturers of the value of good design and commitment to the development of new products. Continental manufacturers are increasing their market shares in traditional markets in addition to other markets. This has led to a high level of imports in some sectors of the industry.

The impression should not be formed that innovation is to do with products only. Chapter 13 has drawn attention to the relatively poor levels of productivity in the British construction industry. Improved production relates to both the buildability and the management of a project. Forms of contract, bidding, communication, planning and monitoring progress are but a few of the activities which lend themselves to new and innovative procedures.

Sources of funding for R&D

Existing sources of funding involve a number of major public sector bodies, (Table 15.5), which include the Department of the Environment (DoE) and Department of Transport (DoT).

Private sector funding, see Table 15.6, is used in-house or is contributed to the established development groups. Other funding organizations include the European Union (EU), the British

Technology Group (BTG), and the World Bank. The BTG has funds available specifically for patentable development.

Apart from the contractors and consultants operating in the construction industry, there is a multiplicity of bodies that contribute publicly and privately to the conduct and operation of R&D; well over one hundred. See Table 15.7.

Although many of the bodies shown in Tables 15.5, 15.6 and 15.7 have a specific area of operation, there is much overlap and duplication and research needs to be better co-ordinated and directed. This is being presently addressed by a number of initiatives and panels including the whole industry Research Strategy Panel. Duplication of

Table 15.5   Existing sources of public sector research funding [9, 13]

- DoE: Several directorates are actively involved in research programmes that concern the construction industry; Construction Sponsorship Directorate; Environmental Protection Group; Planning Directorate; Housing and Urban; Energy Efficient Office. The DoE has an on-line database of DoE and local authority research. The majority of the work required within the DoE is directed to the Building Research Establishment to be undertaken either in-house or by placing external contracts.
- DoT: responsible for all bridge and highways research programmes mainly through competitive tendering; although a substantial number of projects are awarded directly to the Transport Research Laboratory.
- MAFF: work required by the ministry, particularly for coastal protection and flood control.
- MoD: work required by the ministry.
- DTI: supports schemes to promote technology development and best practice in all sections of industry, e.g. Small firms scheme Award for Research & Technology; Support for Projects Under Research; Teaching Company Scheme; New and Renewable Energy; Carrier Technology Programme; Constructional Steelwork.
- ODA: work required by the ODA in support of its policies: primarily at the Building Research Establishment. Transport and Road Research Laboratory. Hydraulics Research Ltd. Institute of Geological Studies.
- NHS: commission research from time to time.
- EPSRC: funded by the DFEE for research which is limited to universities and EPSRC wholly owned institutes
- NERC: geology, oceanography and marine biology, hydrology and so on, mostly at the British Geotechnical Society. Institute of Oceanographic Sciences. Institute of Marine Environmentalists. I. Terrestrial Ecology, etc.
- Local Authorities: *ad hoc* and very limited funding
- British Airports Authority.
- HEFCE.

Table 15.6   Examples of existing sources of private sector research funding [9, 13]

Suppliers

- National Power and Powergen: own laboratories at Leatherhead and Ratcliffe.
- Water Industries mostly for the Water Services Association and in-house research, but some funding available for specific projects required by individual Regional Water Authorities, e.g. reinstatement of roads.
- British Gas: own laboratories at Capenhurst.
- Railways: Railtrack.

Manufacturers

- Brick: mostly in-house or by way of the British Ceramics Research Ltd.
- Cement: in-house or by way of the British Cement Association.
- Timber: by way of the Timber Research and Development Association.
- Paint: mostly in-house or by way of the Paint Research Association.
- Rubber & Plastics: mostly in-house or by way of the Rubber & Plastics Research Association.
- Building Services: mostly in-house or by way of the Building Services Research and Information Association.
- Steel: Steel Construction Institute.

Contractors and consultants

- Some in-house or by way of CIRIA and others: also significant contribution through participation in British Standards Institution committees. Institution of Civil Engineers activities and other bodies.

---

work would not be entirely inappropriate except for the fact that the overall funding of R&D in construction has been shown to be relatively low in comparison with other major industries and many major overseas competitors. With funding thinly spread, the direction of R&D needs to be closely controlled to prevent duplication. There is scope to initiate the co-ordination of R&D effort in construction in co-operation with the major research organizations, both private and public.

Chapter 6 referred to the structure and characteristics of the construction industry which is comprised of relatively few large firms and 190 000 small firms, the latter mainly having under 14 employees, with a low asset base and low margins. It is unlikely that the many small firms would adopt other than traditional techniques or contribute funds to raise investment in R&D, lending little to innovation.

Table 15.7   Other sources of research funding [9, 13]

- EU: funds available from various programmes, e.g. BRITE, EUREKA, etc.; UK could attract much more funding from the EU with a concerted effort: a centre of expertise in guiding applicants through the labyrinth of procedures to secure EU funds is needed. The Construction Industry European Research Club (CIERC) exists to assist companies in all sectors of industry.
- Insurers: generally unwilling to support construction research.
- Property developers: generally unwilling to support construction research.
- BTG (formerly NRDC): funds available for patentable development.
- World Bank: for specific work related to Third World development.
- Wolfson Foundation: rarely for construction projects.

It would seem important, therefore, that the large firms make a greater commitment to innovation, or restructuring takes place in the industry to improve R&D. Perhaps larger firms could combine into bigger groupings, combining their individual expertise, providing greater strength and diversity to prevent Britain falling even further behind its competitors. It is large firms that innovate most and provide leadership to the smaller and medium size firms. The practice over recent years to produce a more fragmented construction industry does not bode well for R&D and the future success of the industry.

More recently there has been a reluctance by government to provide greater funding for basic research or for construction research. The Research Councils have noted 'with dismay' the steady decline in the real value of public funding and stated that increased publicly funded investment is the essential pre-condition for engaging more private sector funds through collaborative programmes. The suggestion that the harnessing of basic scientific research will be fruitful for applied research as construction, like other fields of activity, depends on a wide range of basic scientific research has been dealt with succinctly by de Solla Price [14]:

> Another remarkably widespread wrong idea that has afflicted generations of science policy students holds that science can in some mysterious way be applied to make technology. Quite commonly, it is said that there is a great chain of being that runs from basic science to applied science and thence to development in a natural and orderly progression that takes one from the core of science to technology. Historically, we have almost no examples of an increase in understanding being applied to make new advances in technical competence, but we have many cases of advances in technology

being puzzled out by theoreticians and resulting in the advancement of knowledge. It is not just a clever historical aphorism, but a general truth, that 'thermodynamics owes much more to the steam engine than ever the steam engine owed to thermodynamics.' Again and again, we find new techniques and technologies when one starts by knowing and controlling rather well the know-how without understanding the know-why. We often (but not always) eventually understand how the technique works, and this then leads to modifications and improvements, giving the impression that science and technology run hand in hand. But historically the arrow of causality is largely from the technology to the science.

Conduct of R&D

Most construction professionals are engaged in work involving a careful search and enquiry for information; although this may not entirely constitute research, the discovery of new facts or the collation of known facts by a close study of a subject may be more acceptable as a definition. However worthwhile this work may be to an individual, it is not beneficial to an industry unless it is disseminated, effectively evaluated and incorporated advantageously in construction works. A number of major public sector bodies are involved, through funding and/or undertaking work in construction research and development. These include the Building Research Establishment (BRE) which plays a dominant role in the conduct and dissemination of construction R&D in Britain; the prevention of fire is another major area of investigation at the Establishment. The BRE is part of the DoE. Its main role is to advise the DoE and other government departments. In order to keep advice up to date, the BRE undertakes a broadly-based research programme ranging from the studies of basic properties of construction materials to investigations of specific building problems. Its findings and accumulated expertise are communicated to the industry by means of its information transfer facilities. The BRE has extensive links with research and technical organizations throughout the world. The BRE News, the monthly newsletter in construction, fire and timber editions, reaches a regular audience of 150 000. Advice is given annually in response to over 24 000 enquiries. Employing some 700 staff, the BRE yearly budget is some £34m; 0.07% of the turnover of the industry to which it contributes. Until recently, the BRE has been undertaking £0.5m of work directly for the construction industry but, in future, it is intended that 10–15% of the BRE funding will come from a private sector workload.

Other public sector bodies undertake work in their laboratories largely for their own benefit. Most universities carry out construction research and the value of this has increased from £33m to £54m during the 1990s. Source of funding are the Engineering and Physical Sciences Research Council (EPSRC) and the DFEE through the HEFCE. At present EPSRC fund building and civil engineering by means of a design and technology subcommittee. In the private sector, materials manufacturers, contractors and consultants conduct in-house research and are involved in R&D through established development groups such as the Timber Research and Development Association (TRADA) and the Construction Industry Research and Information Association (CIRIA). One of the private sector organizations, the British Cement Association (BCA) has recently reduced in size and the workforce has fallen from 450 in 1987 to 70 in 1992. Assets have been sold and a small team of technical experts retained to supervise the commissioning of work in other research organizations. This reorganization has shocked the construction industry and opinion suggests that no other major country in Europe will be so badly served in concrete research.

Much, but not all, R&D work undertaken has application to areas of study that are generally acknowledged as problem areas which require investigation or important areas needing further development. It is largely the universities that make a significant contribution to pure research. Although not immediately applicable to existing construction projects, pure research, nonetheless, represents a longer-term investment in the industry. The benefits of longer-term contributions have not been quantified on a scientific basis. The application of discounted cash flow techniques to intellectual investment might startle the protagonists of more generously funded pure research.

Dissemination of the results of R&D

At first sight, selecting and undertaking a piece of research and writing a report giving the results seems a fairly straightforward process. There are many pitfalls with which to contend and this means that often results do not benefit construction practice. Research is not always closely related to what is required by the construction industry; it was pointed out that, during the 1950/60s, the cost of research on shell roofs was a substantial proportion of the total value of shell roofs actually constructed [15]. This is not to say that work which will not quickly and clearly benefit practice should not be undertaken, but that a proper strategy must be used in order to ensure that a balanced portfolio of projects pertinent to industrial requirements is assembled.

When need is established, demonstration and dissemination should be properly planned if the research project is meeting the required performance criteria and indicates useful benefits. If the R&D meets a perceived need by the end user it is likely to be successful. Major users operate on a 'show me' basis. The latter factor was appropriately demonstrated during the successful development of foundation hardware intended for use in offshore construction by a small university based company [16]. During commercial negotiations to licence the hardware, firms would not commit themselves until a strong patent position was established. No doubt many small companies may decide that, without further financial assistance, such a procedure is beyond their resources. A further factor that made the patenting operation more involved and expensive in this case was that, during the initial stage of the research programme, before commercial interests were more closely involved, prior publications and disclosures put some of the work in the public domain. The situation was resolved, but serves to illustrate the conflict and difficulties for the researcher employed in academe who wishes to publicize work for career development and yet retain the degree of secrecy often essential for the innovator in the commercial world. Clearly, the best advice for a researcher who wishes to retain an invention, and prosper from it, is to work on it secretly, patent sensibly, and disclose details only under a confidentiality agreement. This necessitates a close knowledge of patent procedure.

Both public sector and private sector research bodies commit funds to the dissemination of information based on their income from this activity. In addition, there is a multiplicity of journals, conference proceedings and other information which represents a considerable flow of information to the industry. Often the practising builder has little time to read this information or access a database. Yet it is not an isolated incident when a practitioner, requiring more information on a particular subject, consults a relevant specialist paper in a learned journal and finds that it is not clearly evident how the information applies to day-to-day practice. In short, the presentation of the information is not in a form that can be easily used. Some research results are not readily available to the profession representing a waste of research resources. When deciding to place the information in the public domain, research authors need to be more closely pressed by publishers, and at learned society meetings, to indicate clearly the relevance of the subject matter to the profession with clear indications as to its use. Should a key be required to unlock the information, the nature and status of such a key needs to be made clear to practitioners.

As knowledge expands, specialists are confined to narrowing domains whilst generalists attempt to integrate the information sets. The terms 'splitters' and 'lumpers' are used in taxonomy to separate classifiers who like to create new taxa and those who prefer to coalesce categories [17]. Splitters have been in the ascendancy for some time. A more determined effort by the lumpers could do much to collate and interpret existing research results and present them to the industry distilled in such a way as to render the knowledge of increased value. Funding bodies and institutions could encourage more of this work as a recognition that a proper balance of efforts is essential to the development of a coherent intellectual structure.

The value of R&D

In what way has new R&D influenced the construction product? The examples of beneficial research include advanced theoretical techniques such as the finite element method and plastic theory, new design techniques applied to urban drainage; improvements in safety in construction including compressed air working, trenching and the use of chemicals. Changes to Codes of Practice have resulted. These are but a few examples. The CIRIA publication (1985) gives some 60 examples of the benefits of R&D to civil engineering practice alone during the past 25 years. On the other hand, insufficient R&D has given rise to materials shortcomings, particularly with glass reinforced cements, and significant problems with precast buildings. Had the latter problems been dealt with successfully, a potentially satisfactory method of construction which is used elsewhere could have been used in Britain which might have improved productivity and efficiency within the construction industry.

It is interesting to note that R&D is said to reflect techniques rather than artefacts; hence, the reason for fewer construction patents which may be a reflection of the organization of the industry. It has also been suggested that it is a reflection of the nature of the educational process for the construction professional in Britain which often lacks a solid background of technology.

The value of R&D to the construction industry cannot be underestimated. R&D is necessary to maintain international competitiveness and success, particularly as the craft-based traditions of construction diminish and the technological base expands. In particular, civil engineering and building services are technology-based industries which rely increasingly on new materials, new types of plant and equipment. As the construction market shifts periodically across the world, new conditions and constraints relating to environment and materials must be taken into account. This background of constant change and challenges demands an effective R&D base to introduce change effectively and efficiently. Trial and error by the industry is slow and expensive. Clients would be reluctant to accept new and unproven techniques without the reassurance that they have been closely researched to guarantee performance. The cost-benefits of R&D projects are difficult to assess. As an example, perhaps the Ferrybridge disaster might have been avoided (Figure 15.3) and all the consequential remedial work to many existing cooling towers. Relatively small R&D investments in piling and pulverized fuel ash (PFA) in concrete have shown huge benefits and future savings [18]. In the former case, £1m of R&D in 1957–1962 has made possible the construction of real estate to the value of £1.28bn. In the latter case, an initial R&D budget of between £2m and £5m produced savings each year of some £8.75m. If PFA were not used in this way, it would

Figure 15.3   *Was the Ferrybridge disaster a failure of R&D?*

constitute a waste product for which disposal costs would be necessary. Assessments made in 1985 by the Construction Industry Research and Information Association (CIRIA) have shown the cost-benefit of CIRIA's total programme as being well in excess of ten times expenditure, with individual cases providing an annual benefit of 30 times the cost of research. Similar benefits have been reported by individual contractors [9].

Research in the international community

At a recent symposium [11] organized by the European Network of Building Research Institutes (ENBRI) it was acknowledged that the investment in R&D for the construction industry is low; in relation to turnover, it had perhaps the lowest investment in R&D and probably amounted to between half and one per cent of the added value of the construction industry. Several main areas for high priority research projects were identified which would be incorporated into the framework R&D programme. These include:

- a rationalization of performance concepts for buildings, components, equipment and materials; project duration 5–10 years using 250–500 man-years
- integration and rationalization of information management; project duration 4–5 years using 150–200 man-years
- assessment, maintenance and repair of building stock
- low energy, high comfort buildings; project duration 7–10 years
- reduction in raw materials usage
- indoor comfort and health; the core of the project encompassing the design process only would need 50–100 man-years but an extended project, to include the monitoring and control of services and maintenance, could need up to 500 man–years

References

[1] Heyman, J. 'The safety of masonry arches' in *International Journal of Mechanical Sciences*, **2**, 1969
[2] Pavitt, K. *Technical innovation and British economic performance*. Science Policy Research Unit, 1980
[3] Annual review of government funded research and development. 1990
[4] Finniston, M. *Engineering our future*. HMSO. 1980
[5] Annual Reports of the UK Patent Office
[6] *Industrial Property Statistics*. World Intellectual Property Organization

[7] NEDO. *Strategy for construction R&D*. London, 1985
[8] Foster, C. 'Re-group to rebuild: a challenge for the industry'. Duke of Edinburgh Lecture. 1991
[9] *Construction Research and Development*, volumes 1 and 2. Thomas Telford, 1986
[10] UK Construction Sector R&D 1990–1994: review of its funding and provision. Research Focus No. 26. August 1996
[11] *Construction research needs in Europe*. European Network of Building Research Institutions, Luxembourg, 1991
[12] Brookes, A.J. and Stacey, M. 'Discussion paper on component development'. *Design Council/BMIB Seminar*, July 1991
[13] UK Construction Research and Innovation. An overview of opportunities for funding support. DoE, 1995
[14] de Solla Price, D.J. *Little Science, Big Science . . . and Beyond*. Columbia University Press, New York, 1986
[15] McIntosh, J.D. 'Is your research really necessary?' in *Magazine of Concrete Research*, **18**, 1966
[16] Harvey, R.C. and Burley, E. 'Case studies in geotechnics: some expedients for technical innovation' in *Technovation*, Elsevier Science Publishers. Netherlands, 1984
[17] Morowitz, H.J. *The Wine of Life*. Sphere Books, London, 1979
[18] *Research funding in the construction industry*. The Institution of Civil Engineers, 1988

16

Overseas developments

This year alone (and it is not unique) shows the greater financial and technical strength of the foreign firm in the construction industry as a whole.

Sir Monty Finniston, 1990

Background to overseas work

Prior to the First World War, British contractors undertook contracts in many parts of the world, whilst the construction activities of other developed countries were limited largely to their own geographical areas. This included the United States of America whose main preoccupation was developing its own industry and infrastructure. British contractors had the advantage of access to the London money market which made it possible to offer financial facilities for their work. As British financial power diminished in the interwar years, Britain concentrated on a home market which included many of its colonial possessions and this situation continued until the mid 1960s. However, as the traditional markets declined, British companies again competed in the wider world market. Consulting firms were ideally placed to meet the needs of countries which had developed a measure of contracting expertise.

The rapid increase in the prices of oil and gas in 1973 led to an increase in construction projects in countries which benefited from the increased wealth. Many of these countries did not have the trained manpower to meet the needs of the proposed projects and this resulted in an importation of skills from British managing contractors and consultants, in addition to that of labour forces from elsewhere in the world. More recently, the spread of education in developing countries has resulted in a greater part of construction work being undertaken and controlled by these countries. The World Bank has encouraged a greater measure of self-sufficiency as it is advantageous to develop national resources in addition to being cost-effective. Nonetheless, in large and complex works there is still a role for international operators able to mobilize multi-disciplinary project teams and finance or supply specialist services. This can result in joint working agreements or turnkey projects.

A joint working agreement, or joint venture, with a local operator, is often seen as a vehicle for a British consultant or contractor to be involved in overseas contracts, yet subject to less risk than would be the case when operating alone. The involvement can be beset with many pitfalls, not least problems of co-ordination. It is important to define carefully in a written agreement the responsibilities, financial participation and powers of both, or several, parties. A joint venture can also be termed a partnership or consortium. It may be, in effect, a collection of subcontracts with the overall management expenses shared in accordance with the written agreement.

Examples of overseas joint ventures include: (i) the £41.8m contract for the Berkley Bridge in Norfolk, Virginia, undertaken by Tidewater Construction, an American subsidiary of Kier International, with Peter Kiewit of Nebraska, and (ii) the £150m Tate Cairn road tunnel in Hong Kong undertaken by Gammon Construction, owned jointly by Trafalgar House and Jardine Matheson, with Nishinatsu of Japan. Both these examples serve to indicate that the contracts taken on by British companies, or their overseas subsidiaries, represent an important source of international business. However, few British companies are involved in the bulk of the overseas business [1].

The turnkey contract is well known and is a contract in which the promoter allows the contractor freedom to perform his contractual responsibilities. There is a growing practice to allow a measure of technology transfer to the promoter's country.

Land and Marine Engineering, part of the Costain Group, recently carried out an overseas turnkey design and construct contract for India's Oil and Natural Gas Commission. It was the export terminal at Hazira on the west coast of India. The contract was valued at £15m and was substantially funded by the World Bank.

Project finance

The financing of international projects in developing countries is undertaken by several funding organizations and mechanisms. For example, the Multilateral Development Banks which include the World

Bank Group, official development assistance, exports credits and foreign commercial lending. Generally, public sector projects are funded by the banks, but the International Finance Corporation, which is an arm of the World Bank Group, invests funds in private sector projects. Overall, the total capital investment in developing countries in 1983 was £50bn [2, 3]. In addition to this investment, the private sector has played a more important role during the past decade. It is not clear what proportion of the overall capital investment is made in construction but, as over 40% of World Bank investment is on construction projects, this may be a first-stage indicator of the size of the market for construction services. Most developing countries do not seek to employ foreign contractors to the same extent as two decades ago but there are significant opportunities for consultants. This is reflected in the present annual capital value of overseas work in which British contractors and consultants are involved, i.e. £2026m and £44 700m, respectively.

Risk analysis

Contract documents and contractual arrangements are intended to explain the responsibilities of each party to each other. They also define the insurance protection considered necessary. The uncertainties involved from non-commercial risks can vary from delayed completion to political risks. The latter are of greater concern in developing counties but few schemes offer separate coverage for individual political risks. Rather, insurance schemes tend to provide blanket coverage of all risks at a flat premium. Premium rates vary but flat premiums tend to range between 0.5% and 1.5% of the insured amount [3]. Although these costs may be borne initially by the promoter, the user of the project, which is invariably the community, is responsible for funding the insurance premiums.

Work done by contractors

British contractors, or their subsidiaries, undertook work overseas in 1994 to the value of £3.20bn at current prices. This represents over 5% of all work carried out in Britain and overseas. Table 16.1 and Figure 16.1 show the value of the overseas work carried out in the past two decades and indicates the areas of the world in which the work was undertaken. Also shown is overseas work as a percentage of total output, home and overseas, although it is not strictly comparable as it is not in accordance with the Standard Industrial Classification. The British contractors world market share is shown which, although not completely reliable, indicates trends.

Table 16.1 Overseas construction activity by contractors [4, 5]

	70/71	71/72	72/73	73/74	74/75	75/76	76/77	77/78	78/79	79/80	80/81	81/82	82/83	83/84	84/85	85/86	86	87	88	89	90	91	92	93	94
	£m												£m												
European Community	10	8	8	8	12	12	5	8	15	34	37	38	24	21	45	79	91	73	39	40	59	120	137	178	293
Rest of Europe	17	25	43	46	51	64	56	109	91	83	105	17	80	58	57	23	21	32	42	37	81	59	44	75	63
Middle East in Asia	43	62	52	58	144	373	590	788	818	630	467	521	669	675	700	585	422	230	111	113	109	220	262	347	318
Middle East in Africa	1	1	1	–	1	1	22	27	33	61	66	50	62	66	48	81	92	55	64	63	25	{	{	{	{
Hong Kong																125	71	57	54	57	214	141	190	409	567
Rest of Asia	22	28	23	16	16	16	37	46	79	83	96	97	264	306	306	136	205	116	58	83	102	120	202	287	343
Rest of Africa	65	69	73	115	126	233	336	398	421	234	200	304	524	557	482	381	330	199	152	162	219	180	215	183	233
Americas	75	87	63	62	52	84	112	127	124	167	204	339	436	440	611	594	480	565	731	813	957	1016	1062	1060	1143
Oceania	36	37	49	57	55	74	92	94	97	92	96	112	255	217	237	199	237	248	105	159	260	310	222	227	227
Total	269	317	312	362	457	857	1250	1597	1678	1384	1271	1478	2314	2340	2486	2203	1949	1575	1356	1527	2026*	2166	2334	2766	3187
% of all work (home and overseas)	4.3	4.7	4.4	4.0	4.5	7.2	9.3	11.0	9.7	6.8	5.5	4.3	9.3	8.8	8.2	6.9	5.7	4.2	2.9	2.8	3.8	4.1	4.7	5.3	5.3
% World market share [5]	na	na	na	na	na	na	na	na	na	na	na	6.8	6.1	6.8	6.8	na	9.5	10.7	10.0	12.8	10.4	na	na	na	na

Note: the brace ({) indicates Middle East in Asia and Middle East in Africa combined for the years 91–94 (220, 262, 347, 318).

* provisional

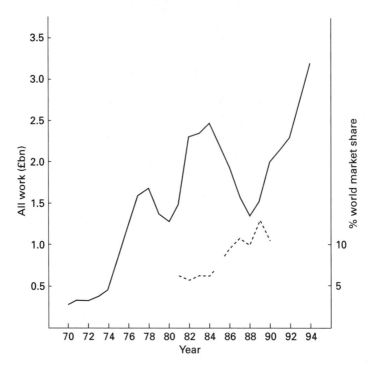

Figure 16.1   *Overseas construction activity by contractors*

It is not possible to accurately represent Table 16.1 in real prices as inflation is a variable and uncertain factor throughout the world. Measured at current prices the amount of work increased steadily from 1970 to 1978 before declining. A peak in workload occurred again in 1984 followed by a further decline. During the past few years the value of work has doubled from approximately £1.5bn to £3bn per year. Overall, 90% of the work has been accounted for by ten British companies. The increase in work throughout the 1970s was due to the increase in contracts in the Middle East. The overall decline after 1978 was linked to the decline in this work although the influence of inflation in Britain, not experienced by competitors, may have affected the ability to win contracts, albeit marginally. Since the early 1980s the value of work done has increased substantially by a better performance in the Americas. By 1988, over 50% of the overseas work done was in the Americas. The work in the far east has increased despite the increase in strength of major international competitors in Japan and Korea. Much of the increase has been in Hong Kong. The six biggest companies in Japan are on average four times larger than the biggest of Britain's top six contractors. This is shown in Table 16.2.

Also shown in the table is Costain Group plc. Both Kumgai Gumi Co. Ltd. and Costain Group plc undertake approximately 50% of their work overseas; the Kumgai Gumi Co. Ltd. is over five times bigger than the Costain group plc. In comparison, the top twenty Japanese contractors are each bigger than the largest British contractor.

British contractors still have a significant amount of work in their traditional markets in Asia and Africa. However, the value of overseas work done by British contractors in Europe, including the European Union, although increasing is relatively small.

The value of overseas work, expressed as a percentage of all work carried out by British contractors, has been less than 10% apart from the late 1970s. If British contractors sought overseas work due to a decline in the home market, they were successful apart from in the early 1980s. In terms of world market share based on the orders received by the top 250 firms, the success of British contractors has held up well in recent years [5]. The market share has risen from 4.8% in 1980/81 to 10.4% in 1990. Britain performs about as well as some other European countries, e.g. France and Italy, but is an 'also ran' when compared with Japan and the United States. It should be noted that reference [5], from which this information is obtained, is not confined to construction only but includes other work such as process engineering. Therefore, Table 16.1 should be used with caution. There is a measure of uncertainty regarding the overseas market for construction services.

Table 16.2   Total sales of the biggest Japanese companies [5]

	1985 Contracts ($m US)	
	Total	Overseas
Kajima Corporation	4953	12.7%
Shimizu Construction Co. Ltd	4779	12.7%
Kumgai Gumi Co. Ltd	4692	46.4%
Taisei Corporation	4641	5.7%
Ohbayashi Corporation	4280	8.8%
Takenaka Komuten Co. Ltd	3787	5.0%
Costain Group plc	845	49.9%

Work done by consultants

Table 16.3 and Figure 16.2 show the capital value at current prices of the overseas construction projects undertaken by British consulting engineers who are members of the Association of Consulting Engineers. Also shown is the consultants' world market share based on the value of the work billed by international design firms. This is not an exhaustive list as work is also undertaken by non-members. For the reasons given earlier, the value of the work has increased significantly since 1973 but slowed down during the 1980s. The value of the work in 1990 was less than 1982 despite the contribution of the Channel Tunnel in 1989 and 1990. Without this contribution the current value in 1990 would be much the same as that in 1981. There has been a highly encouraging performance since 1991 and that work entrusted to members has very nearly doubled.

The work is not listed in accordance with the Standard Industrial Classification and includes the work of consulting engineers undertaking electrical and mechanical engineering projects linked to construction work. The range of work includes land planning and development, transportation projects, commercial and industrial buildings, water supply and sewerage works. These are the largest areas of work. Projects are undertaken in all areas of the world with the largest proportions being carried out in the Far East, Africa, Europe and the Middle East. Since 1990, the work in the middle east has increased from £6725m to £16 145m; in Africa it has declined slightly from £9282m to £8695m, in the Far East due to Hong Kong being a major source of work and with China investing large sums in development of industry and infrastructure, work has increased substantially from £14 264m to £35 096m; in Europe there has been an increase from £8458m to £12 117m. The value of projects in the Americas is relatively small. This contrasts with British contractors whose influence in the Far East has declined whilst that in the Americas has increased significantly.

Table 16.4 compares the capital cost of the work in hand in 1990 and in 1995 by project and region. British consulting engineers are involved in construction projects throughout the world which have a capital value about 30% in excess of the total construction work carried out in Britain. Nearly 150 firms are listed, although this includes overseas practices; there are some 50 distinct firms listed by the Association of Consulting Engineers. They are involved in approximately 140 countries. Figure 16.3 shows an example of work entrusted to British consultants. Over 40% of the work is carried out in the Far East.

Table 16.3 Capital value of work undertaken by engineering consultants [6]

	Year										
	70	71	72	73	74	75	76	77	78	79	80
Total work £bn	1.9	2.2	3.2	4.1	4.5	7.5	14.7	25.4	31.8	36.0	35.7%
% World market share [5]	na	na	na	na	na	na	na	na	na	na	15.3

Year															
	81	82	83	84	85	86	87	88	89	90	91	92	93	94	95
Total work £bn	38.5	46.2	45.0	46.5	52.5	38.5	31.4	33.5	34.3*	44.5*	41.8	49.0	65.3	63.9	81.7
% World market share [5]	12.8	13.8	15.4	13.1	na	13.6	na	10.6	15.9	17.4	na	na	na	na	na

* Includes Channel Tunnel contracts with non-British agencies.

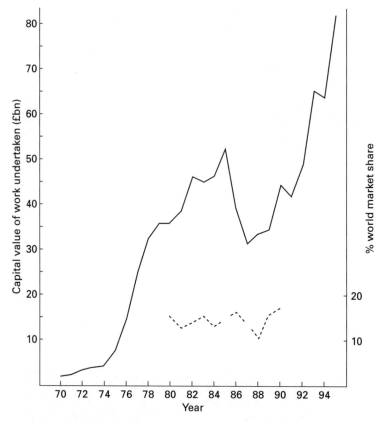

Figure 16.2   *Capital value of work undertaken by engineering consultants*

Land planning and development contributed 20% and roads, bridges and tunnels over 11% of the total workload. Structural commercial work and water supply also make substantial contribution. During 1995 the consultants earnings overseas amounted to £730m. This is a significant contribution to the United Kingdom's invisible earnings.

In contrast to the contractors who tend to carry out the smaller part of their work overseas, some of the consultants have as little as 10% of their turnover in Britain with, overall, as much as 75% of the work overseas. This has been facilitated by the adoption of joint ventures with local consultants. Financial assistance is provided by the Multilateral Development Banks or, nearer home, the Department of Trade. The Export Credits Guarantee Department insures consultants for fees, financing bonds and other export risks.

Work done by surveyors

The overseas potential of the surveying profession has been strongly promoted by the Royal Institution of Chartered Surveyors (RICS). Organizations involved in the promotions have included the Overseas Trade Services, British Consultants Bureau and the International Federation of Surveyors. The RICS has organized, and been involved in, visits, seminars and workshops in many overseas countries including Vietnam, India and Germany [7].

The members overseas foreign currency earnings have not increased significantly since 1984 (Table 16.5). In real terms the earnings have probably decreased. The estimated earnings for 1991 are £132m which is the highest figure since surveys began in 1979 [7].

Contribution to balance of payments

Table 16.5 and Figure 16.4 show the contribution to the United Kingdom balance of payments made by contractors, consulting engineers, architects/quantity surveyors and chartered surveyors. (The table reflects the way in which the information has been collected in

Figure 16.3 *This 200 m high platinum mine chimney is an example of work entrusted to British consultants*

Table 16.4 Capital cost of work in hand in 1990 and 1995 £m [6]

	Airports		Chemical Petroleum Gas plants		Desalination		Drainage Sewerage Refuse		Electrical Mechanical Services		Harbours Docks Sea Defence		Hydro-Electric		Irrigation		Land Planning Development		Nuclear Power Stations	
	1990	1995	1990	1995	1990	1995	1990	1995	1990	1995	1990	1995	1990	1995	1990	1995	1990	1995	1990	1995
Middle East	4	491	—	211	171	—	438	476	185	138	415	262	—	—	10	26	88	2 342	—	—
Africa	197	55	1	498	21	—	559	456	60	31	136	45	626	217	154	27	161	310	—	—
Indian sub-continent	15	487	—	361	—	—	144	383	5	54	129	169	113	489	683	509	3	15	—	—
Far East, Hong Kong China, Japan and Korea	1200	2326	—	135	—	—	1159	2044	478	85	357	1457	340	—	188	21	1047	12 893	40	—
Australia and New Zealand	—	—	—	10	—	—	—	—	50	—	20	13	—	—	—	—	—	—	—	—
North and Central America and West Indies	4	510	—	8	—	—	157	21	30	356	16	6	1	40	—	—	1	72	92	—
South America	1	—	—	1	—	—	—	—	1	—	1	17	—	—	11	—	—	—	—	—
EU	25	68	—	1617	—	—	—	161	30	582	251	64	40	—	—	—	—	221	—	—
Rest of Europe	—	241	—	505	—	—	—	111	39	—	20	1394	—	—	—	—	—	95	—	—
Total	1446	4178	1	3346	192	—	2457	3652	878	1246	1345	3427	1120	746	1046	583	1300	15 948	132	—

Table 16.4 Cont.

	Railways		Roads Bridges Tunnels		Structural Commercial		Structural Industrial		Thermal Power Stations		Transmission of Power		Water Supply		Miscellaneous		Total	
	1990	1995	1990	1995	1990	1995	1990	1995	1990	1995	1990	1995	1990	1995	1990	1995	1990	1995
Middle East	—	—	1761	1515	535	301	399	361	2489	1476	169	166	33	1509	28	6871	6 725	16 145
Africa	62	—	844	681	281	77	418	189	97	59	128	30	5512	5708	25	12	9 282	8 695
Indian sub-continent	—	771	432	431	77	1 717	12	358	28	—	278	34	496	104	10	50	2 425	5 932
Far East, Hong Kong China, Japan and Korea	415	4191	3565	4787	2934	4 677	196	320	561	1250	472	239	989	249	323	422	14 264	35 096
Australia and New Zealand	53	1	352	8	1337	—	28	—	57	—	204	—	—	—	4	—	2 181	32
North and Central America and West Indies	—	532	84	20	415	1 364	—	13	21	2	10	—	42	12	—	6	873	2 922
South America	—	1700	10	4	—	—	1	179	—	320	450	24	—	—	2	117	477	702
EU	5500		719	1545	164	1 749	37	327	—	56	—	—	—	8	1110	560	7 876	8 658
Rest of Europe	—	180	—	216	113	119	90	330	94	113	49	—	173	40	4	125	582	3 469
Total	6030	7375	7767	9207	5856	10 004	1181	2077	3347	3276	1760	493	7245	7630	1506	8163	44 685	81 651

Table 16.5 Contribution of construction to the UK balance of payments [8]

	Year										
	70	71	72	73	74	75	76	77	78	79	80
Contractors (£m)	43	52	58	71	94	130	174	201	213	151	126
Consultants (£m)	35	48	65	84	108	136	214	305	370	401	423
Architects/ Quantity Surveyors (£m)	6	8	11	14	18	23	36	51	63	64	79
Chartered Surveyors (£m)	na	na	na	na	na	na	na	na	na	54	59

	Year														
	81	82	83	84	85	86	87	88	89	90	91	92	93	94	95
Contractors (£m)	124	126	132	na	na	na	na	na	na	na	na	na	na	na	na
Consultants (£m)	487	565	561	577	562	508	418	400	440	480	450	625	605	631	730
Architects/ Quantity Surveyors (£m)	88	98	103	na	na	na	na	na	na	na	na	na	na	na	na
Chartered Surveyors (£m)	65	73	74	92	99	110	97	100	110	96	na	na	na	na	na

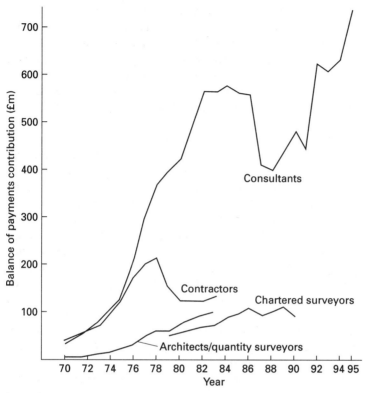

Figure 16.4   *Contribution of construction to the UK balance of payments*

recent years.) Since 1971 the contribution of the consultants has
increased by an order of magnitude whilst that of the contractors has
only doubled. Since the early 1980s the contributions have remained
fairly steady at current prices. In real terms this is a significant
decrease. Nonetheless, the construction industry is providing a much
needed contribution to Britain's trade.

Construction in the European Union

The five largest construction markets in Europe in terms of output are
Britain, France, Italy, Spain and Germany. Their relative sizes are
shown in Table 1.9 in Chapter 1.

In terms of output, Germany remains the dominant country. Britain
is distinguished in terms of the small housing and public sector
markets; this suggests relatively poor housing and infrastructure

compared to our European counterparts. The retention of a strong public sector market in mainland Europe ensures that the construction industries are not fully exposed to the vagaries of the private sector. They are not subject to a frequent stop/go policy which has a debilitating influence on the construction industries. In contrast, the commercial, i.e. private sector, market in Britain appears better developed. Apart from Spain, Britain spends the lowest per capita sum on construction.

France, due to size and dispersed population, has developed a strong construction industry. Emphasis is placed on housing and the civil engineering sectors. The latter is related to transportation and is dominated by the public sector. France takes the Single Market seriously and has a policy for infrastructure development aimed at positioning France at the core of the European transport systems. The country's energy needs are met largely by nuclear energy. French contractors achieve a high turnover and, in comparison with Britain, have lower assets [9]. The ten largest construction companies in France account for 35% of the total turnover. Many small companies are active in the regions. Both consultants and contractors operate in a highly professional manner and foreign firms find it difficult to operate profitably in France. The traditional markets for French contractors are declining and they are diversifying and undertaking work elsewhere in Europe, including Britain.

In Italy, the structure of the industry is similar to France in that a few large firms dominate and many small firms operate in the regions. Although few foreign companies undertake work in Italy, there has been a number of joint venture agreements and the smaller contractors welcome relationships with foreign contractors to allow them to tender for large contracts. Larger contractors are looking for more overseas markets to replace the dwindling traditional markets at home, North Africa and the Middle East. In contrast, Spain is improving its infrastructure, particularly the road and rail network, and Spanish contractors are busy with the home market. The private sector has increased in importance. Restructuring of the industry is taking place to meet future competition.

Germany is currently investing less in new housing but, overall, housing investment has remained fairly constant during the past few years. This sum includes both new and renovation work and is comparable to the overall investment that Britain makes in construction. The main development areas are in roads, commercial buildings and waste treatment. Foreign firms may achieve market penetration by means of joint venture agreements or establishing a German subsidiary by means of a takeover. German contractors are looking for opportunities outside Germany and have carried out work in Europe and the United States. They place emphasis on quality and reliability and set high standards [10]. The unification of West and East Germany may stretch the capabilities of the consulting services in West Germany and opportunities could be available for foreign firms with established skills.

Although there has been a dramatic increase in work for British consultants in Europe during the past two years, it may not be the pattern for the future. There is the prospect of greater competition in Europe as the mainland European firms seek contracts outside their home territories. Firms that can satisfy both the demand for technical expertise and organize funding will be more likely to succeed. In particular, France, Italy and Germany will be major competitors.

References

[1] British Business. Construction Overseas. July 1989
[2] OECD/IMF/MDB/World Bank annual report
[3] Minch, D.B. 'Project finance – the multilateral development banks'. *Proceedings of a Conference on Management of International Construction Projects*. The Institution of Civil Engineers. November 1984
[4] Department of the Environment. *Housing and construction statistics*. HMSO, London
[5] *Engineering News Record*. 1970–1995
[6] *International work entrusted to members 1980–1995*. Association of Consulting Engineers.
[7] *International progress report to the General Council 1991/92*. Royal Institution of Chartered Surveyors.
[8] Central Statistical office. *Annual Abstract of Statistics*. HMSO, London
[9] Murray, S. 'Structure of the European Construction Industry – an Overview'. *European Construction Industry Analysis Conference*. London, 1989
[10] Engineering Consultancy in the European Community: BRITAIN; FRANCE; ITALY; SPAIN; WEST GERMANY; Association of Consulting Engineers/Foreningen af Raadgwende Ingenioret/Orde van Nederlandse Raadgevende Ingenieurs. Thomas Telford, London, 1990

Abbreviations and acronyms

ABE	Association of Building Engineers
ACE	Association of Consulting Engineers
ADR	Alternative Dispute Resolution
APTC	Administrative, Professional, Technical, and Clerical
ARCUK	Architects Registration Council of the United Kingdom
ASCE	American Society of Civil Engineers
ASI	Architects and Surveyors Institute
BA	Bachelor of Arts
BCA	British Cement Association
BCIS	Building Cost Information Service
BEC	Building Employers Confederation
BIAT	British Institute of Architectural Technicians
BIM	British Institute of Management
BPF	British Property Federation
BRE	Building Research Establishment
BRITE	Basic Research in Industrial Technologies for Europe
BS	British Standards
BSc	Bachelor of Science
BTEC	Business and Technology Education Council
BTG	British Technology Group
CAD	Computer Aided Design
CBCS	Chartered Building Company Scheme
CBI	Confederation of British Industry
CCPI	Co-ordinated Committee for Project Information
CESMM	Civil Engineering Standard Method of Measurement
CGLI	City and Guilds of London Institute
CIArb	Chartered Institute of Arbitrators
CIBSE	Chartered Institute of Building Services Engineers

CIC	Construction Industry Council
CID	Construction Industry Directorate
CIEC	Construction Industry Employers Council
CIERC	Construction Industry European Research Club
CIOB	Chartered Institute of Building
CIPS	Chartered Institute of Purchasing and Supply
CIRIA	Construction Industry Research and Information Association
CISC	Construction Industry Standing Conference
CITB	Construction Industry Training Board
CNAA	Council for National Academic Awards
CPD	Continuing Professional Development
CSO	Central Statistical Office
DEn	Department of Energy
DFEE	Department for Education and Employment
DH	Department of Health
DoE	Department of the Environment
DoT	Department of Transport
DSS	Department of Social Security
DTI	Department of Trade and Industry
EETPU	Electrical, Electronic, Telecommunications and Plumbing Union
ENBRI	European Network of Building Research Institutes
EPSRC	Engineering and Physical Sciences Research Council
ESF	European Social Fund
ETA	European Technical Approval
EU	European Union
EUREKA	European Research and Co-ordination Agency
FCEC	Federation of Civil Engineering Contractors
FE	Further Education
FEFC	Further Education Funding Council
FIDIC	Federation Internationale des Ingenieurs Conseils
FM	Facilities Management
FT	Financial Times
FTAT	Furniture, Timber and Allied Trades Union
GB	Great Britain
GCSE	General Certificate of Secondary Education
GDP	Gross Domestic Product
GMBATU	General, Municipal, Boilermakers and Allied Trades Union
GNP	Gross National Product
GNVQ	General National Vocational Qualification
GP	General Practice

HE	Higher Education
HEFC	Higher Education Funding Councils, e.g. HEFCE (Higher Education Funding Council for England)
HMC	Household Mortgage Corporation
HMSO	Her Majesty's Stationery Office
HNC	Higher National Certificate
HND	Higher National Diploma
HSE	Health and Safety Executive
IBCO	Institute of Building Control Offices
ICE	Institution of Civil Engineers
ICES	Institution of Civil Engineering Surveyors
IFC	International Form of Contract
IMF	International Monetary Fund
IMI	Inovative Manufacturing Initiative
ISO	International Organisation for Standardisation
IStructE	Institution of Structural Engineers
ISVA	Institute of Surveyors, Valuers and Auctioneers
IT	Information Technology
IU	Intensity of Use
JCT	Joint Contracts Tribunal
MAFF	Ministry of Agriculture, Fisheries and Food
MCG	Major Contractor's Group
MDB	Multi-national Development Bank
MIT	Massachusetts Institute of Technology
MoD	Ministry of Defence
MPhil	Master of Philosophy
MSc	Master of Science
NCG	National Contractors Group
NCVQ	National Council for Vocational Qualifications
NEC	New Engineering Contract
NEDO	National Economic Development Office
NERC	Natural Environmental Research Council
NHS	National Health Service
NJCC	National Joint Consultative Council
NRDC	National Research and Development Council
NSCC	National Specialist Contractors Council
NVQ	National Vocational Qualification
NWR	National Working Rule
ODA	Overseas Development Agency
OECD	Organisation for Economic Co-operation and Development
OSO	Offshore Supplies Office
PAYE	Pay as You Earn

PERT	Project Evaluation and Review Technique
PFA	Pulverised Fuel Ash
PFI	Private Finance Initiative
PhD	Doctor of Philosophy
PLC	Public Limited Company
PSA	Property Service Agency
QS	Quantity Surveyor
R&D	Research and Development
REX	Robotic Excavator
RIAS	Royal Incorporation of Architects in Scotland
RIBA	Royal Institute of British Architects
RICS	Royal Institution of Chartered Surveyors
RTPI	Royal Town Planning Institute
SCI	Steel Construction Institute
SIC	Standard Industrial Classification
SECG	Specialist Engineering Contractors Group
SMM	Standard Method of Measurement
TEC	Training and Enterprise Council
TGWU	Transport and General Workers Union
TRADA	Timber Research and Development Association
UCATT	Union of Construction, Allied Trades and Technicians
UK	United Kingdom
UFC	Universities Funding Council
USA	United States of America
VAT	Value Added Tax
YTS	Youth Training Scheme

Index